Date Due

BRODART, INC. Cat. No. 23 233 Printed in U.S.A.

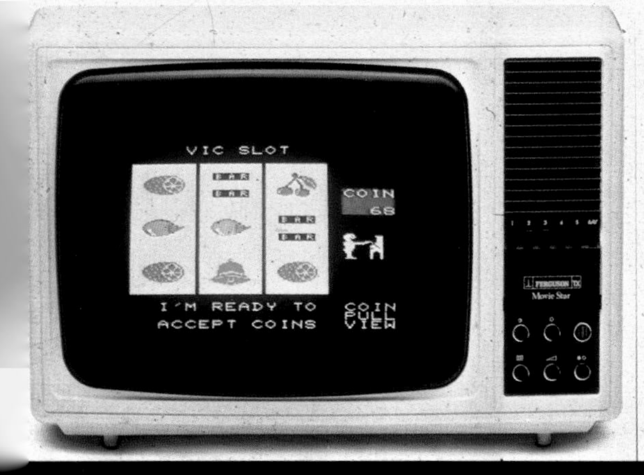
VIC SLOT
BAR
COIN
68
I'M READY TO
ACCEPT COINS
COIN
PULL
VIEW

1982 May
Loan Tot 28000.00
Payment 347.81
Interest 338.33
Tot 338.33
Principal 9.47
Tot 9.47
Rem 27990.53
Press 'E' to end or
any key to continue

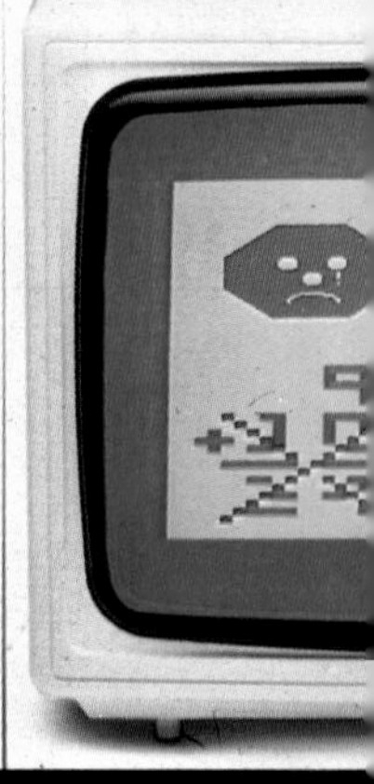

COINS 71 BETS
ROYAL FLUSH
STRAIGHT FLUSH
FOUR OF A KIND
FULL HOUSE
FLUSH
STRAIGHT
THREE OF A KIND
TWO PAIRS
JACKS OR BETTER
CANCEL RETURN
1 2 3 4 5

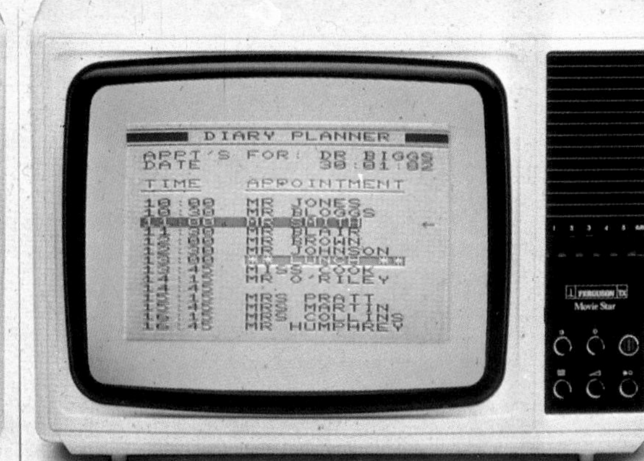
DIARY PLANNER
APPT'S FOR: DR BIGGS
DATE 30:01:82
TIME APPOINTMENT
10:00 MR JONES
10:30 MR BLOGGS
MR BLAIR
MR BROWN
MR JOHNSON
MRS COOK
MR O'RILEY
MRS PRATT
MRS MARTIN
MRS COLLINS
16:40 MR HUMPHREY

PERIOD
BOOKING

DECIMALS
26.74
+26.27

x2
x10
x5
SCORE
FUEL
HI

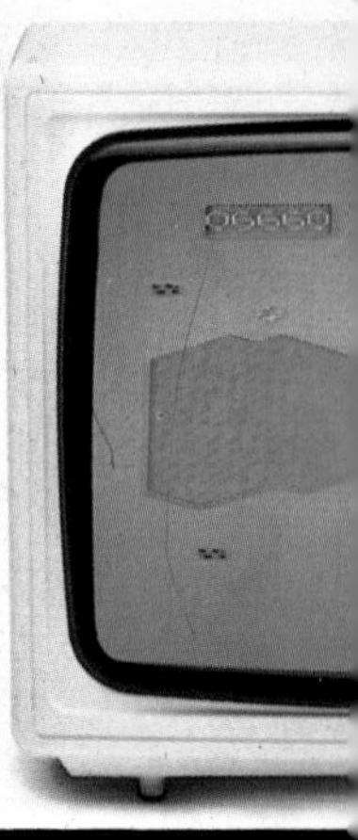

I MOVE B5/A4
A B C D E F G H
YOUR MOVE?

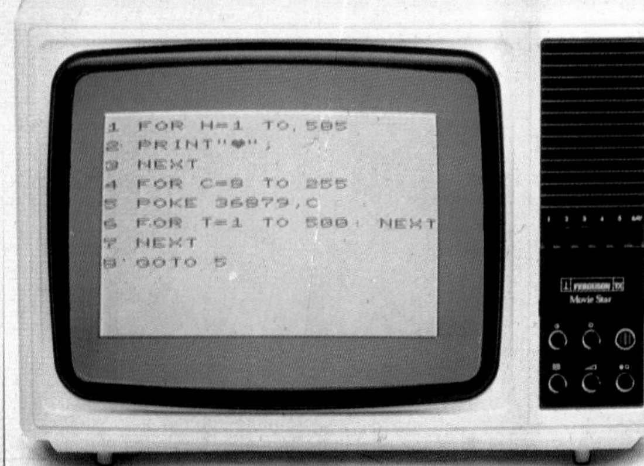
1 FOR H=1 TO 505
2 PRINT"♥";
3 NEXT
4 FOR C=8 TO 255
5 POKE 36879,C
6 FOR T=1 TO 500: NEXT
7 NEXT
8 GOTO 5

SCORE 0 5000 3

1-UP HI-SCORE 2-UP
7420 7420
BONUS 1000 x1
BALL

VIC HOME ACCOUNTS
ITEM NO. 15
GAS BILL
AMOUNT £ 52.36
DATE REC'D 24:01:82
PAID ? Y/N Y
CHEQUE NO.? 0532321
PRESS F1 FOR NEXT ITEM

SCORE 24 BEST 43

PIE CHART
40%
10%
30%
20%

--- VIC MAILING LIST ---
Name |Smith
Initials |J.P.
Number/Name |134
Street |High Street
Town |Slough
County |Berks
Postcode |SL1 4BG
Tel. |55160
STD |0753

CHEMISTRY QUIZ
SYMBOL NA
S*DIUM
D G I K M
R S T U
LETTERS USED
NEXT GUESS? K

Designers Malcolm Smythe
Ben White
Art Director Charles Matheson
Editor James McCarter
Researcher Dee Robinson
Consultant Richard Johnson

Illustrators Denis Bishop
Paul Cooper
Chris Forsey
Hayward Art Group
Nigel Partridge
Roger Phillips

First published 1982 by
William Collins Sons & Co Ltd

Published in the U.S.A. in 1983 by Gloucester Press

Printed in Belgium

ISBN 531-04584-6

Library of Congress
Catalog Card Number 82-50856

Contents

The Inside Story

VIDEO

Gareth Renowden

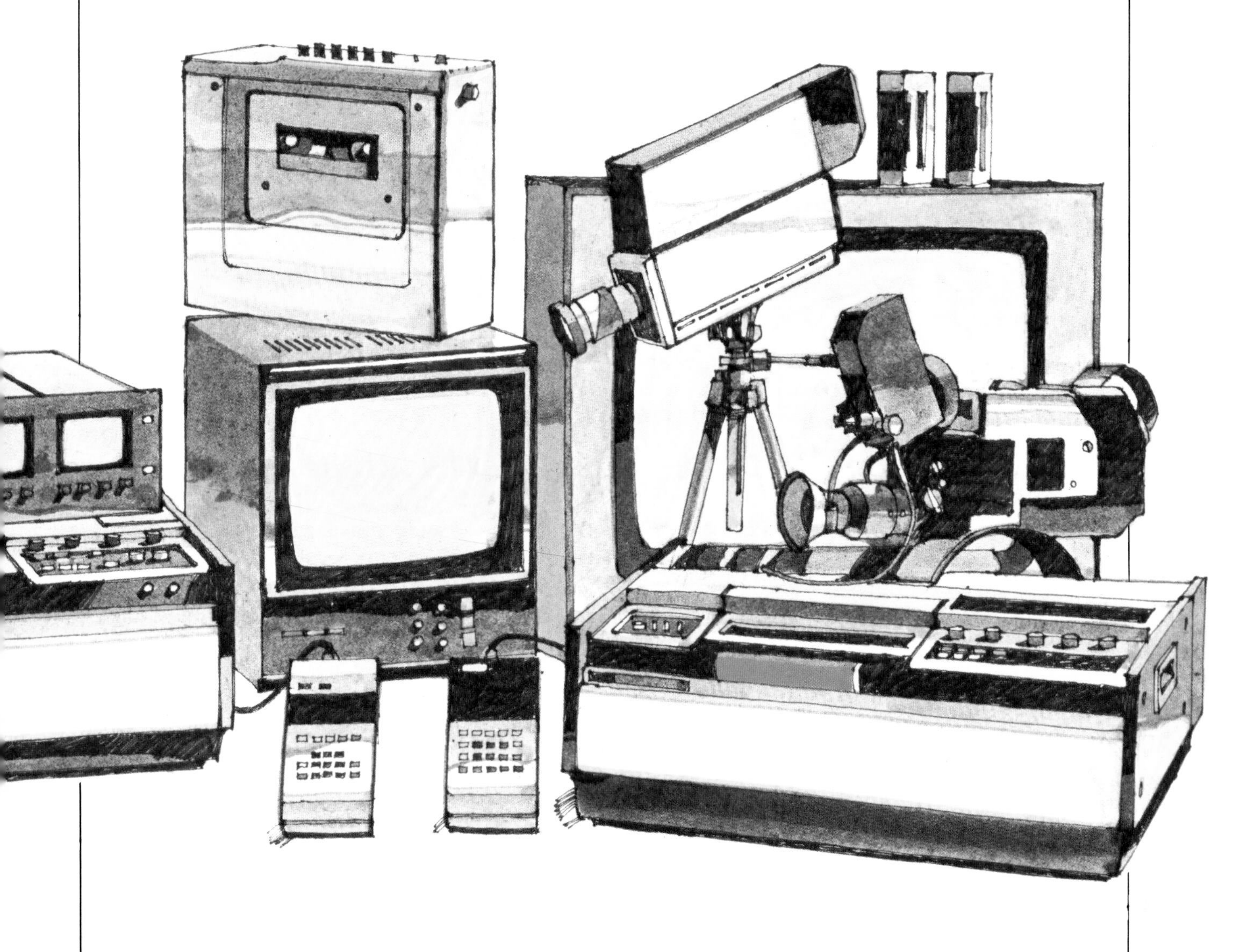

GLOUCESTER PRESS
NEW YORK · TORONTO · 1983

What Does Video Mean?

In today's world, we are surrounded by video. Virtually every shopping center and main street has shops packed with the latest video equipment, either for sale or hire. With television, video tapes and video recorders, and electronic games such as Space Invaders, most of us use video every day of our lives. With satellites, cables and TV broadcasts, video spans the world, and has even sent back "live" pictures from the surface of the Moon.

When people think of video, they tend to think only of the feature films available on video cassette, and the video recorder used in the home. But video has a much greater scope than that. Any picture that is created by electronic means and shown on a TV screen is part of the video world. The pictures shown can range from a simple home movie shot with a video camera, to sophisticated computer-created images.

Video has come a long way since television was first developed in the 1920s and 30s. Today's video systems, thanks to modern microtechnology, have a huge range of applications. With magnetic video tape, images can now be recorded and instantly replayed, giving much greater flexibility than is possible with conventional film, which needs processing before it can be shown.

In schools, offices, hospitals and factories, and in the home, video is used to provide entertainment and education, and to help make work easier and more efficient. And as microchip technology becomes more and more sophisticated, the range of video's applications can only increase over the next ten to twenty years.

Illustrated here are just some of the applications of video. In the factory (1), technicians monitor what is happening on the production line, using video screens. Video games in amusement arcades (2) are a popular pastime. In the TV studio (3), all kinds of sophisticated video equipment is used to create TV broadcasts. Video is also used in hospitals (4), both as a tool to help surgeons and doctors and as a teaching aid to help medical students. And, more and more, video is being used in schools and colleges (5). Video has made news gathering faster and more efficient, so that interviews taken on the street (6), can be relayed direct to the TV company's newsroom (7).

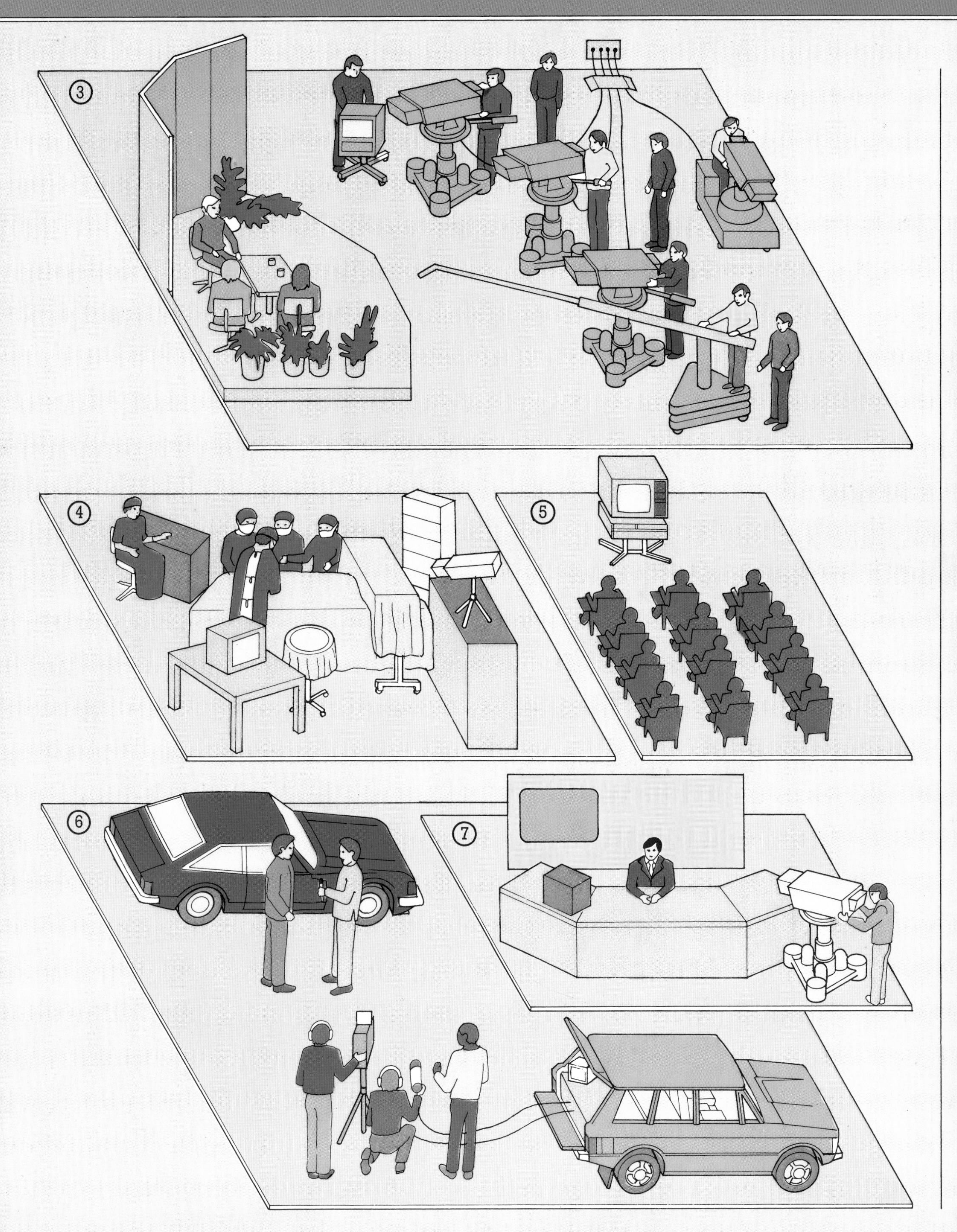
③
④
⑤
⑥
⑦

Video in the Home

This household is well equipped with the latest electronic gadgetry, all of it relying on video in one way or another. At the moment, there are probably few homes like this one, but all the equipment shown is already for sale.

The father is sitting at the home computer, doing the household accounts, or perhaps some office work. If he needs to, he can link up with his company's computer using the telephone. The home computer can also be used to learn a foreign language, book theater tickets or even to order the week's shopping. Every piece of information is shown on the computer-linked TV screen.

Across the room, two other members of the family are playing a video game on the main TV set. There are hundreds of different games to choose from. The family's video recorder and library of video tapes are stored beneath the TV. Some are pre-recorded feature films, either bought outright or rented for a short period. Others are favorite TV programs, which may have been automatically recorded on blank video tapes by the video recorder, even while the family was away from home.

The whole scene is being recorded using a video camera and portable video recorder. Later on, the recording can be edited using the two video recorders. The recording can be shown as a home movie on TV, or even sent to friends or relatives as a video "letter."

The video system at home

Home video equipment
1 Color TV receiver
2 Video tape recorder
3 Pre-recorded video tapes
4 Computer video game
5 Video game cassettes
6 Video camera
7 Portable video tape recorder
8 Blank cassettes
9 Color TV receiver
10 Video tape recorder
11 Home computer

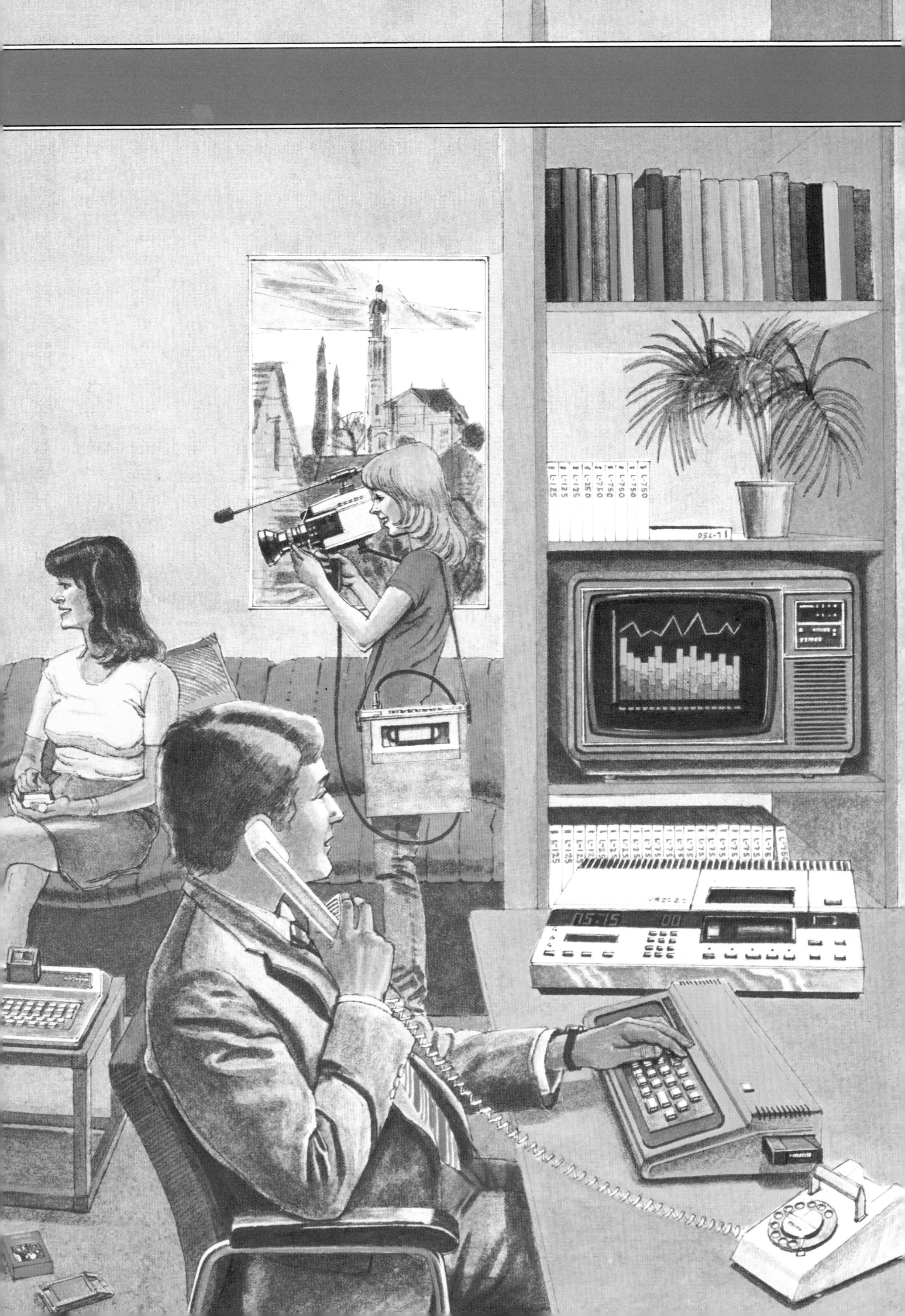
L-125
L-250
L-750

The Video System

In a complete home video system, the TV set (1) acts as the display unit for information provided from a wide range of equipment. Each part of the system is simply plugged into the set. A home computer (2) can then display the information stored in its memory. And with a link to a public computer (3) – systems such as AT&T's videotex – the amount of information instantly available becomes enormous. Plug in a video game (4) – in effect a simple computer programmed to play the game and follow the player's instructions – and the TV screen becomes an intergalactic battleground! With a portable video recorder (5) and camera (6) you can make home movies that can be shown instantly on the TV screen. And, of course, the recorder can also tape TV programs on blank video tape cassettes, or show pre-recorded video feature films (7). To complete the system, all that's needed is a special effects unit (8), and a link-up with the hi-fi (9), to edit simple home movies and give them a stereo soundtrack.

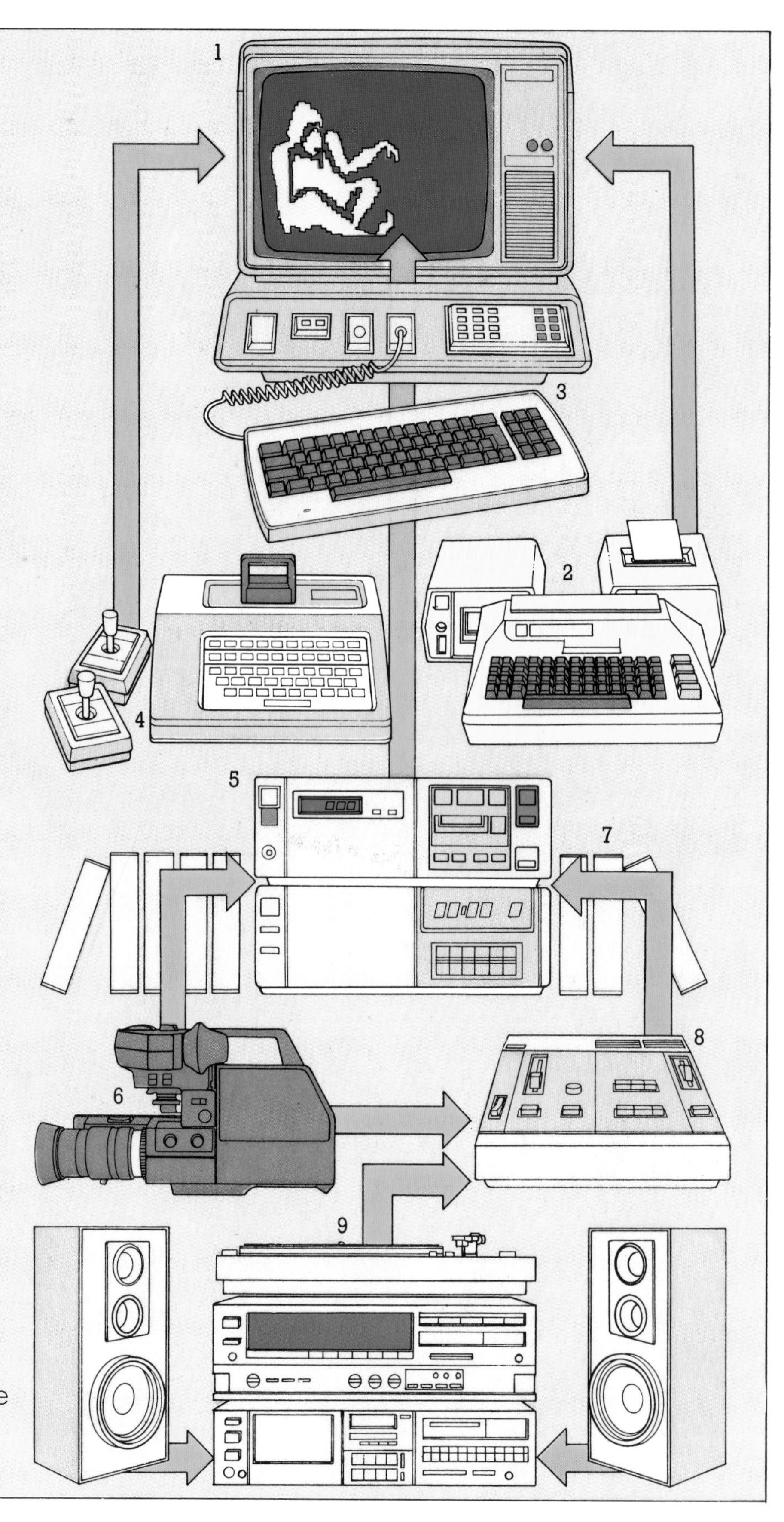

Visual Display

The TV screen on which all video information is displayed is the front of a "cathode ray tube." The tube is made of glass, and has had all the air removed from inside it to create a vacuum. At the narrow end of the tube is the electron gun (1). When a high voltage is passed through this, charged particles called electrons, are given off. These electrons are electro-magnetically focused into beams (2). They then pass through charged metal plates (3), which can aim the beams to any part of the TV screen. The screen is covered by millions of phosphor dots, laid down in colored strips of red, green and blue (4). Each of the three electron beams is aimed at a particular color dot. When the dots are hit by a beam, they give off light to form one small part of the TV picture. The "shadow mask" (5) ensures that each beam hits the right color dot, to get the correct color effect. Various combinations and levels of brightness of the three colors can create a lifelike color picture. To make up the complete picture, the beams sweep, or "scan," across the dots on the screen in a series of horizontal lines. Some color TVs have three separate guns (6) to produce the three electron beams, but in the most modern system the three beams are produced by only one electron gun (7).

The cathode ray tube

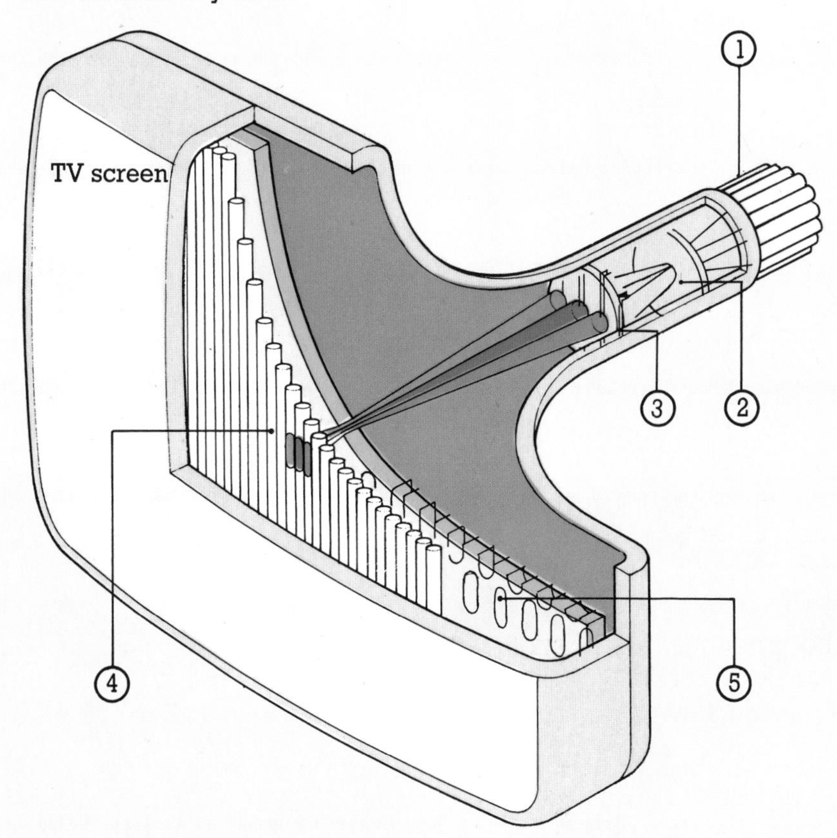

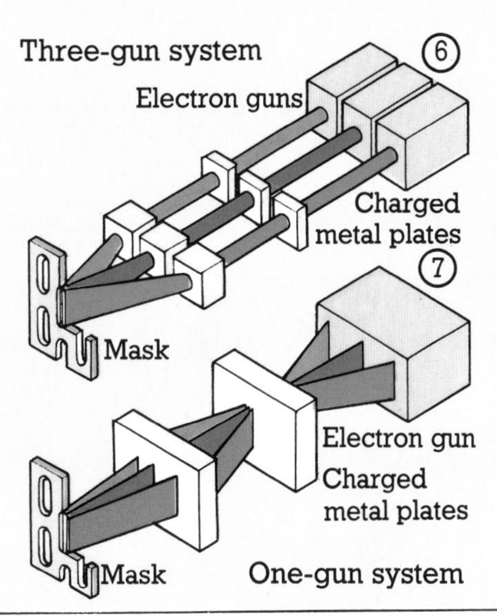

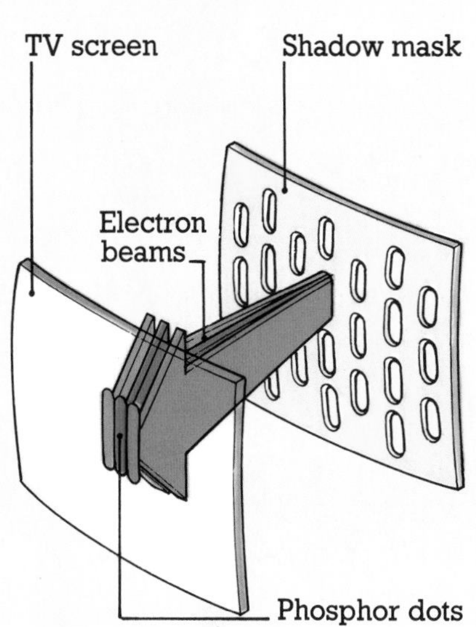

Building the Image

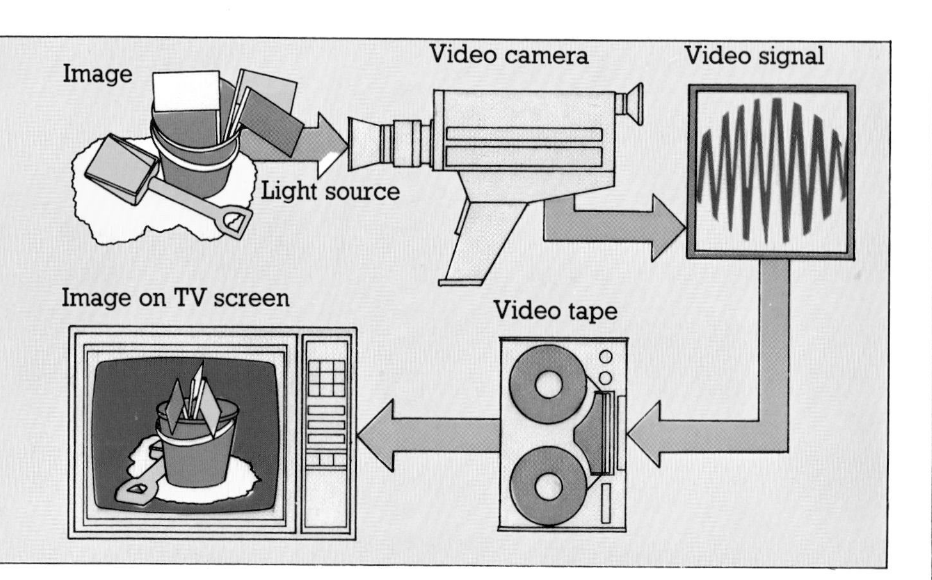

In a video camera, the light from an image creates an electrical video signal. This can be recorded on magnetic tape for replay later, or sent direct to the TV monitor.

Video Signal

The video signal consists of changes in the amount of electrical energy passed down a wire. These changes can be shown as waves. The simplest signal (1) is a regular change from maximum to minimum energy.

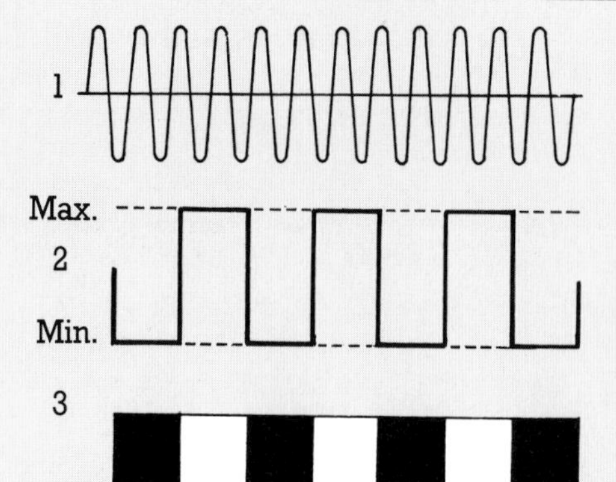

The greater the electrical energy of the video signal, the greater the intensity of the light given off by the phosphor dots on the TV screen. No energy means no light is given off at all. A signal rising from zero to maximum energy and back in a series of steps (2), would result in a series of black-white-black flashes from the phosphor dots (3). But to carry all the information that makes up a TV picture, the video signal has to be much more complicated than this.

In effect, video signals are codes, and the TV receiver is the code-breaker. The signal shown contains the information that makes up one line of a TV picture. First, a pulse (1) tells the TV to begin the line. Then comes a "color burst" (2). This tells the receiver how to separate the different color parts – red, green and blue – contained in the signal.

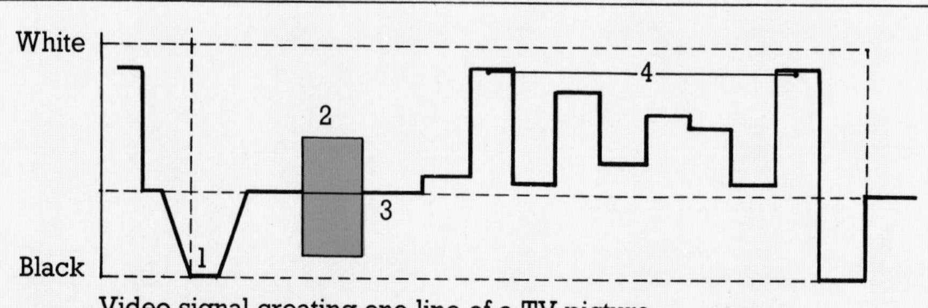

Video signal creating one line of a TV picture

Next, the "black level" (3) tells the TV the lowest level of light to be given off in the line. The rest of the signal (4) consists of changes in electrical energy, which the TV decodes to create the changes in light intensity along the line.

Raster Scan

The way that the electron beams sweep across the screen in a series of lines, to create a picture, is called "raster" scanning. In the U.K., the picture is made up of 625 lines, in the U.S., 525 lines. The beams scan the screen in one direction, to make up half the image, and then "fill in" the other half by scanning down again in the same direction. Each line is made in 0.000064 seconds, and 30 frames ("images") are made each second – so the scanning happens far too fast to be detected by the human eye.

"Raster" scan

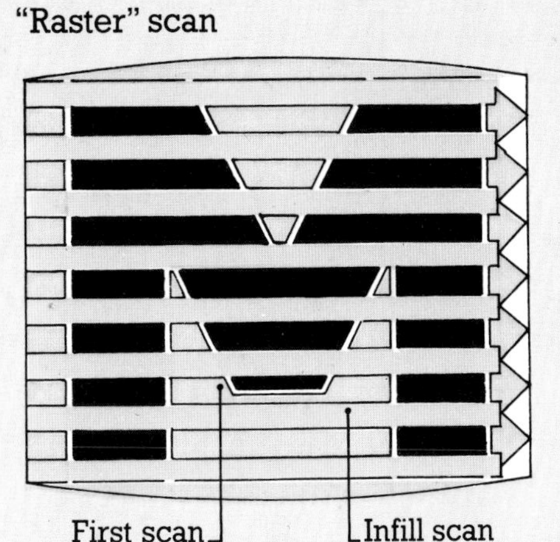

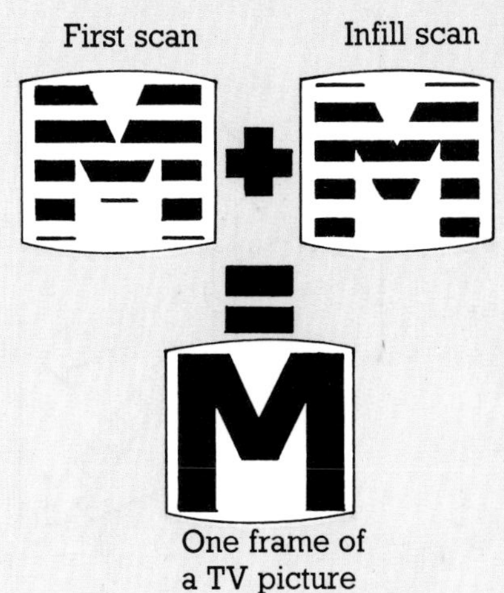

One frame of a TV picture

Quadrascan

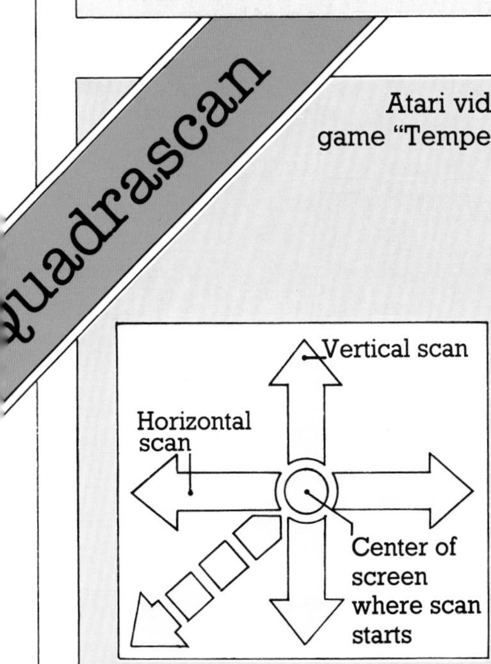

Scan deflected from center

Atari video game "Tempest"

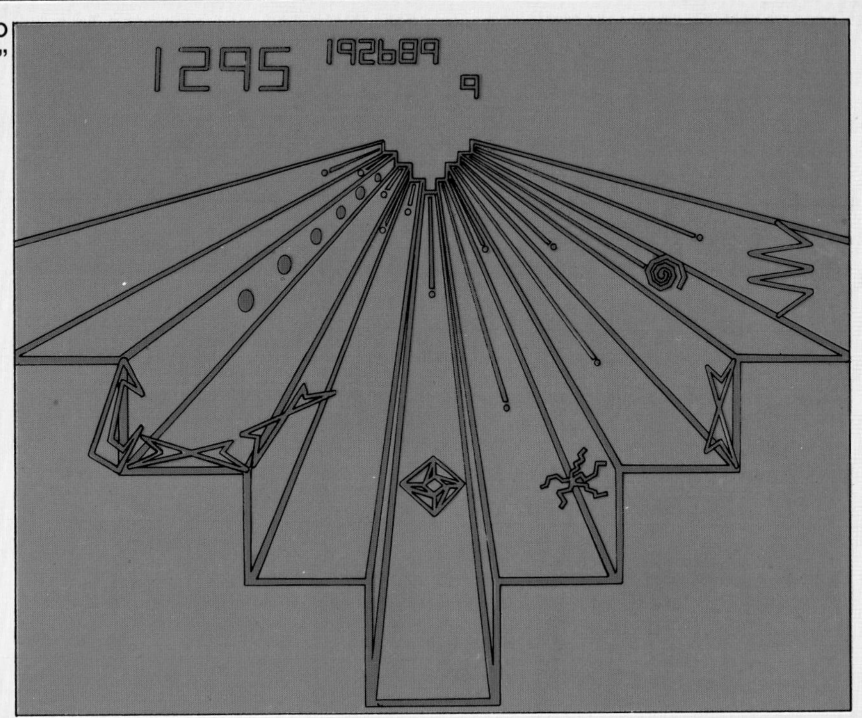

Another method of scanning is called Quadrascan. In this, the image is created by lines deflected from the center of the screen, rather than lines moving across it (above). Quadrascan is used for computer-generated images and also in some video games, like the one shown above. The images produced by Quadrascan are very clearly defined, and stunning three-dimensional effects can be achieved, to create exciting and fast-moving video games.

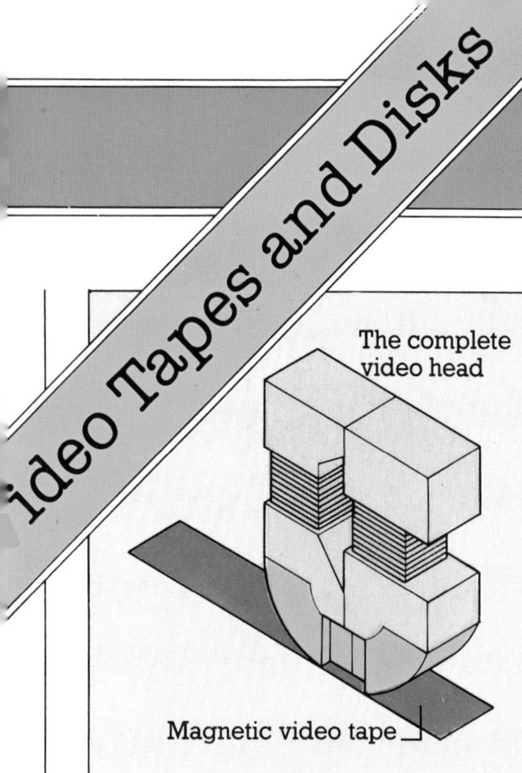

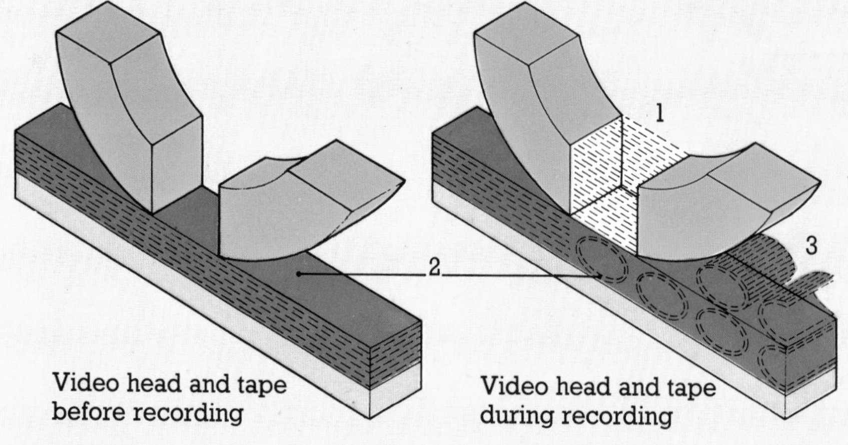

Video signals are electrical, and this means that they can be recorded on magnetic tape. To do this, the signal is processed and then used to create a magnetic field at the video head (1) in the video recorder. The tape, with its magnetic coating (2), is passed over the heads, and the magnetic field there creates a corresponding magnetic field on the tape (3). Each recorder has two heads, mounted on a rapidly spinning drum (4). To pack as much information as possible onto the tape, each head in turn lays down the magnetic field in a series of thin, angled stripes (5). The tape is automatically guided through the video recorder by a series of guide rollers. The soundtrack (6) that accompanies the pictures is laid down as a separate magnetic band at the top of the video tape, by the audio head (7). To replay the video tape, the whole process is reversed. The magnetic field on the tape creates a magnetic field at the tape heads. The recorder converts this field back into an electrical signal which creates the image on the TV screen.

Video tape recording

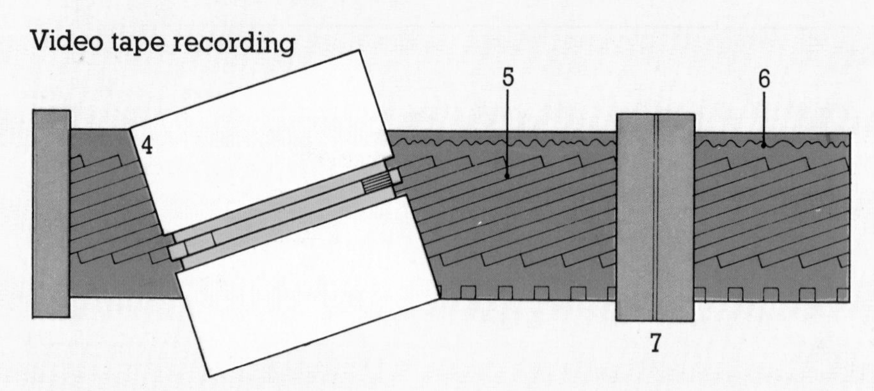

The tape path

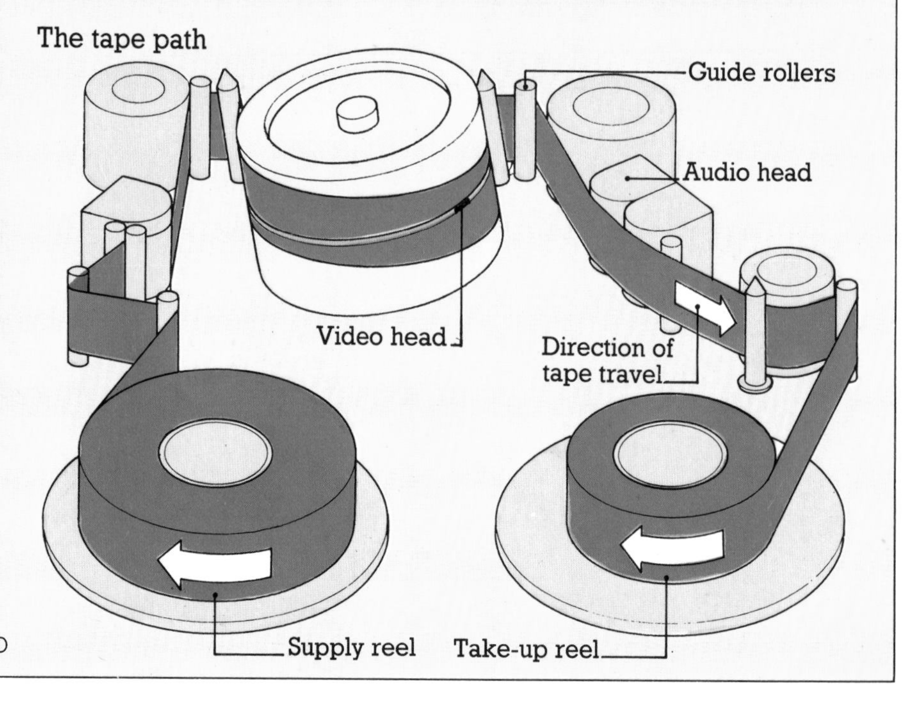

Video Disk 1

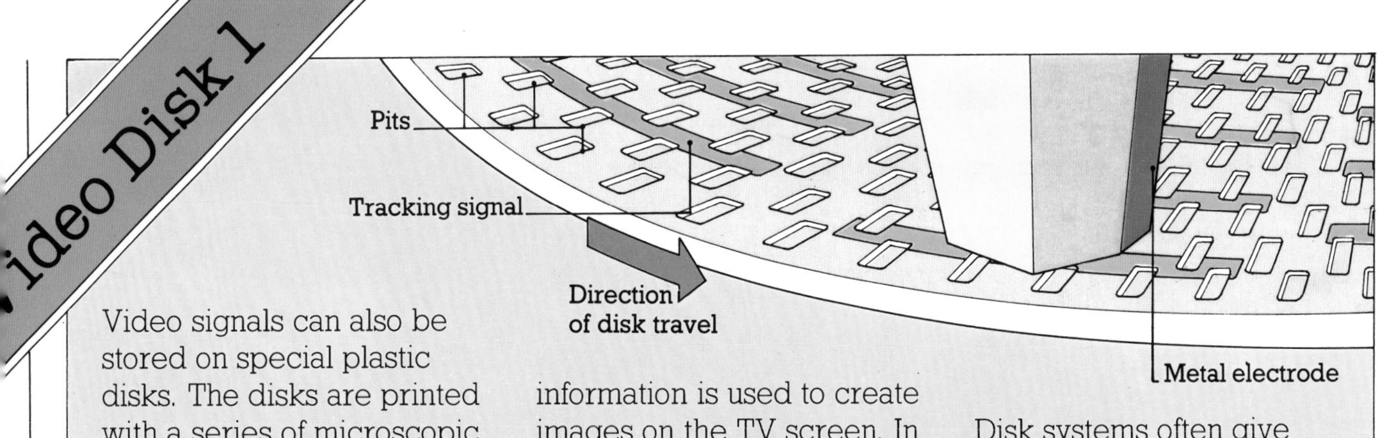

Video signals can also be stored on special plastic disks. The disks are printed with a series of microscopic pits. The arrangement of these pits can be "sensed" by a metal electrode, and this information is used to create images on the TV screen. In the system shown above, special tracking lines guide the electrode over the pits.

Disk systems often give better quality pictures than tapes, but they cannot be used to record in the home.

Video Disk 2

One of the most exciting and sophisticated disk, systems on the market today actually uses a small laser beam to read the information on the video disk. This system, known as LaserVision, has the video information printed on the disk as pits beneath a protective layer. The replay unit's laser beam is directed onto the pits on the disk through a series of mirrors and prisms (below). The relfected beam "tells" the machine the arrangement of the pits, and this information is converted into an electrical signal to create the image on the TV screen.

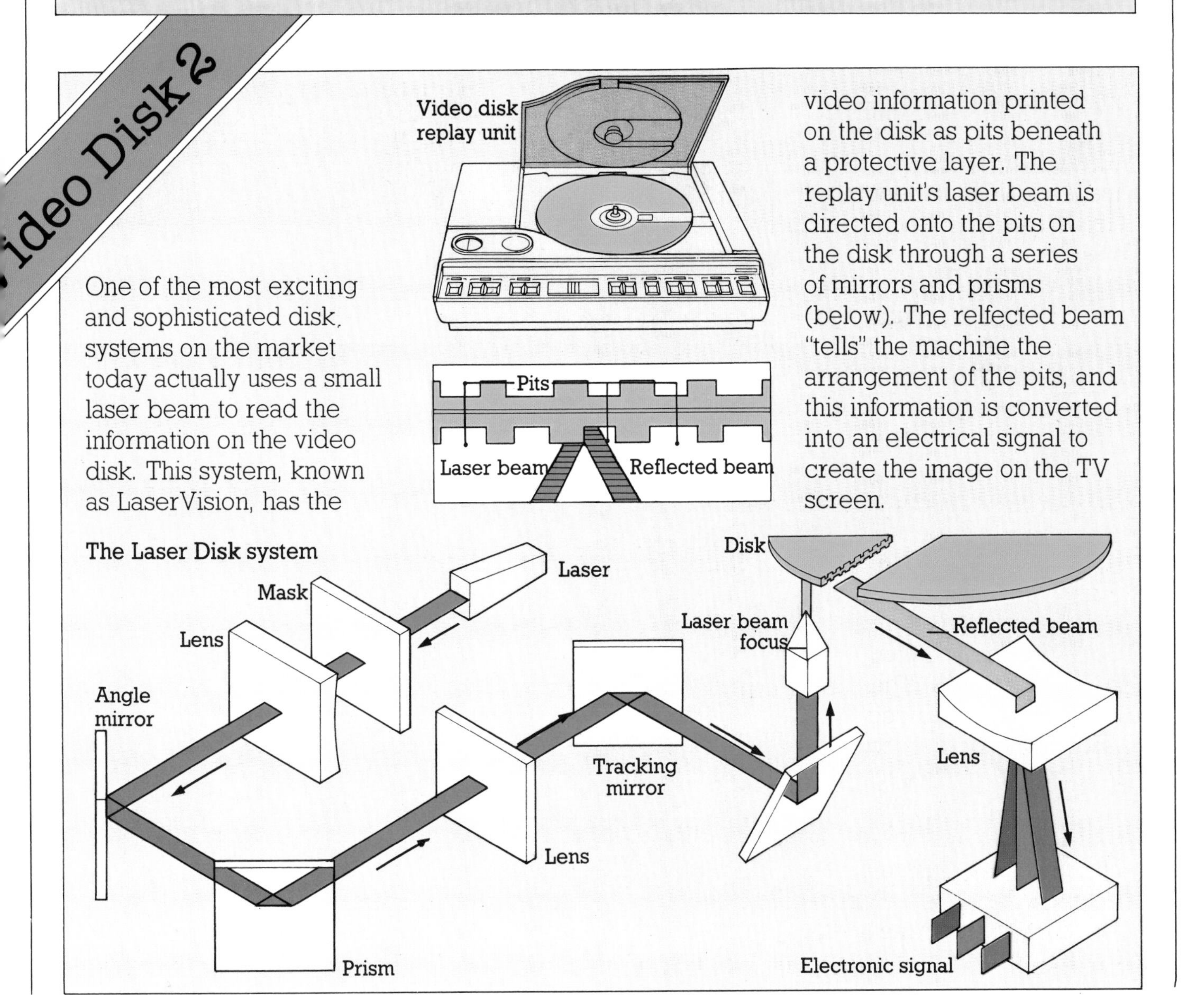

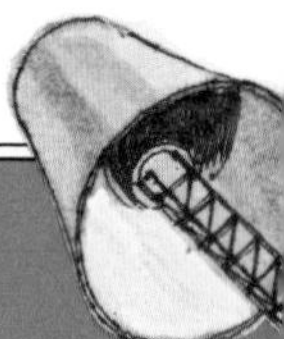

Satellite Broadcasting

How is it possible for TV stations to show "live" pictures from the other side of the world? The answer lies in space, with one of the many communications satellites orbiting Earth. TV broadcasts travel only in straight lines, but they move at the speed of light, and can travel great distances. The pictures being taken of, for example, the world heavyweight boxing championship fight in Las Vegas, Nevada, are specially processed and beamed up to a satellite. The satellite then "bounces" them back to Earth for viewers in London. The whole process takes less than a second, so the viewers can see the action virtually as it is happening.

Sports and news programs are the biggest users of satellite broadcasts. The pictures shown on the evening news program may have been taken only a few hours earlier, thousands of miles away. With satellites, people can be given up-to-the-minute information on all the important events going on in the world.

But satellites have other, equally important uses. The satellite shown here is the Indian National Satellite (INSAT), which combines a number of functions. It will help with communications to remote parts of the massive Indian subcontinent, and, by keeping a 24-hour observation on the weather, it can give advance warning of possible climatic disasters, such as typhoons or floods.

INSAT will also relay TV broadcasts, and broadcast direct to community TV sets in rural villages. The Indian government hopes to use these direct broadcasts as part of a nationwide educational and development project, with programs on health care, more efficient agricultural techniques and general literacy, to try to improve the country's standard of living.

The location and distribution of INSAT receiving stations on the Indian subcontinent.

- ■ Large receiving stations
- ● Medium receiving stations
- ▲ Remote-area receiving stations

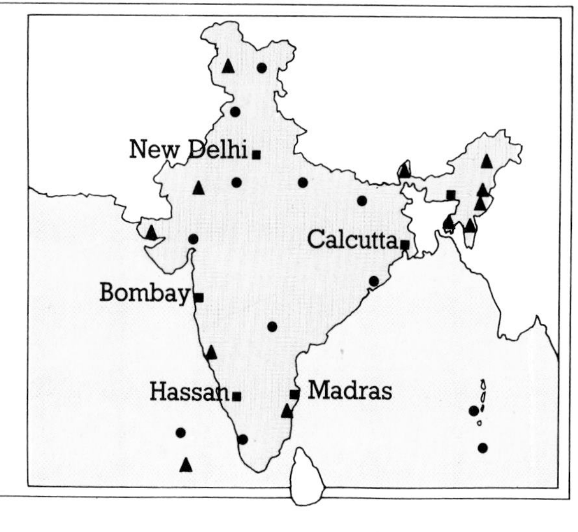

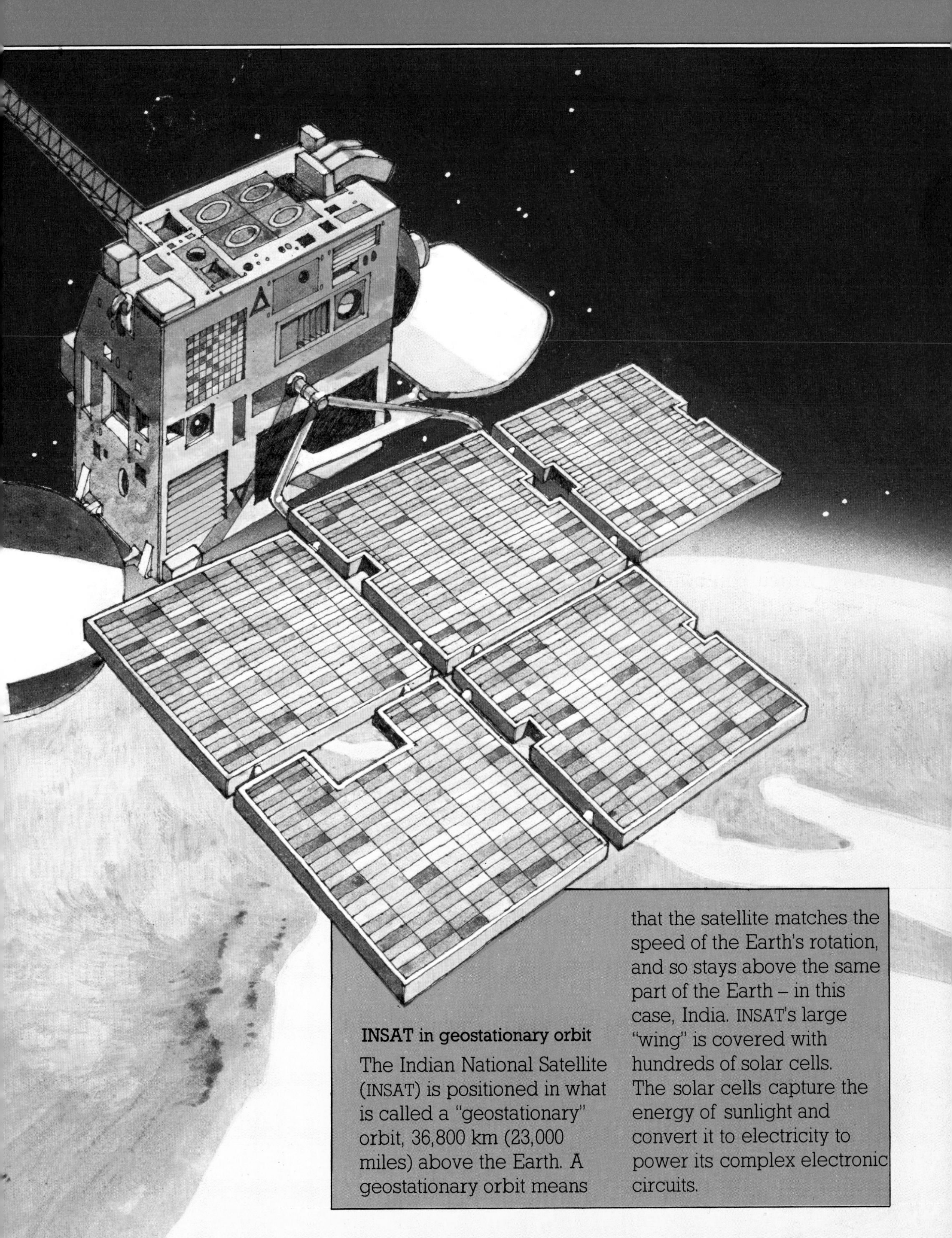

INSAT in geostationary orbit

The Indian National Satellite (INSAT) is positioned in what is called a "geostationary" orbit, 36,800 km (23,000 miles) above the Earth. A geostationary orbit means that the satellite matches the speed of the Earth's rotation, and so stays above the same part of the Earth – in this case, India. INSAT's large "wing" is covered with hundreds of solar cells. The solar cells capture the energy of sunlight and convert it to electricity to power its complex electronic circuits.

Broadcasting

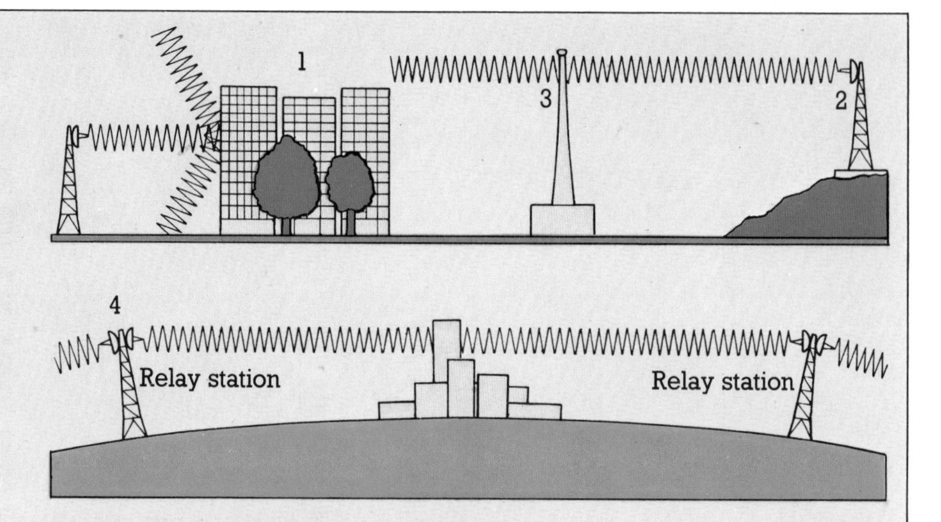

TV broadcasts by the major TV companies are the commonest way that video signals reach the home. Because TV signals travel only in straight lines, they can easily be obstructed by hills or tall buildings (1). So the TV companies build their transmitters on high ground (2) and use relay stations (3) to make sure that their broadcasts reach as many people as possible. The curvature of the Earth's surface also causes problems for TV broadcasts. To overcome this, TV companies build a series of relay stations (4). These TV transmitters and relay stations are some of the world's tallest structures, rising, in some cases, to over 600 m (1,970 ft). Building a lot of such transmitters, in order to be able to broadcast over an entire country, is very expensive. It is cheaper to send a communications satellite into orbit above the Earth's atmosphere (5). This can then beam down TV broadcasts over a wide area. The broadcasts are picked up using special dish antennas, and then fed to the TV companies.

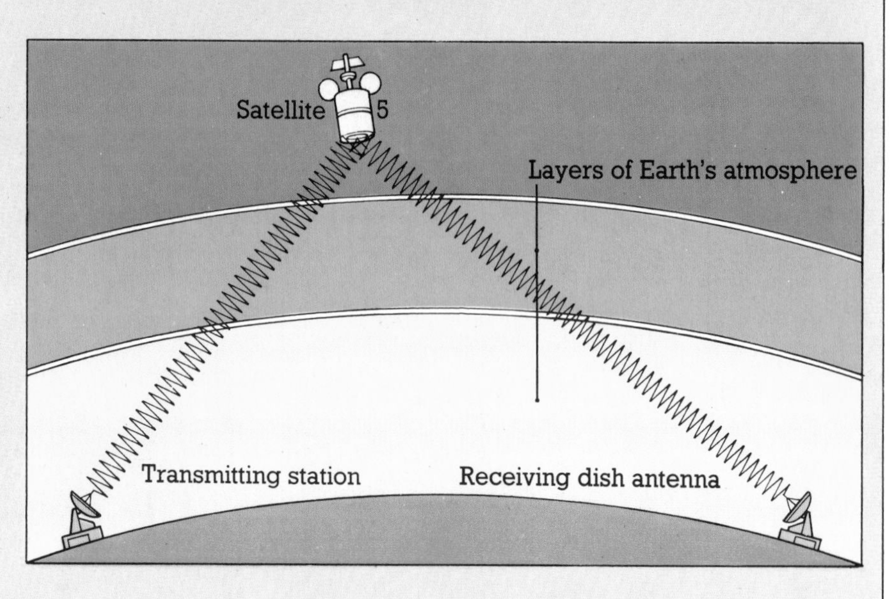

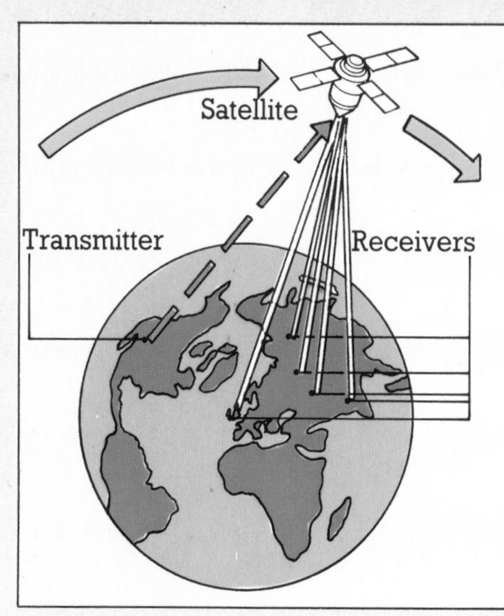

Not all communications satellites are placed in a geostationary orbit. Some revolve around the Earth on a fixed flight path. This means that they can relay TV broadcasts to different parts of the world as they pass overhead. The signal beamed up to the satellite from the TV station has to be accurately timed to the satellite's passage.

Cable Relay

Video signals can also be sent using special cables. A central cable TV station puts programs together, often by just replaying video tapes, with a "video jockey" providing the links between separate items. The signal from the tapes is amplified – made more powerful – and then fed into the cable network. The latest networks use "fiber-optic" cables, capable of carrying over 100 different programs at the same time. This makes cable TV ideal for showing special programs to small audiences, such as local farmers, businesses, schools or hospitals, as well as for home viewing.

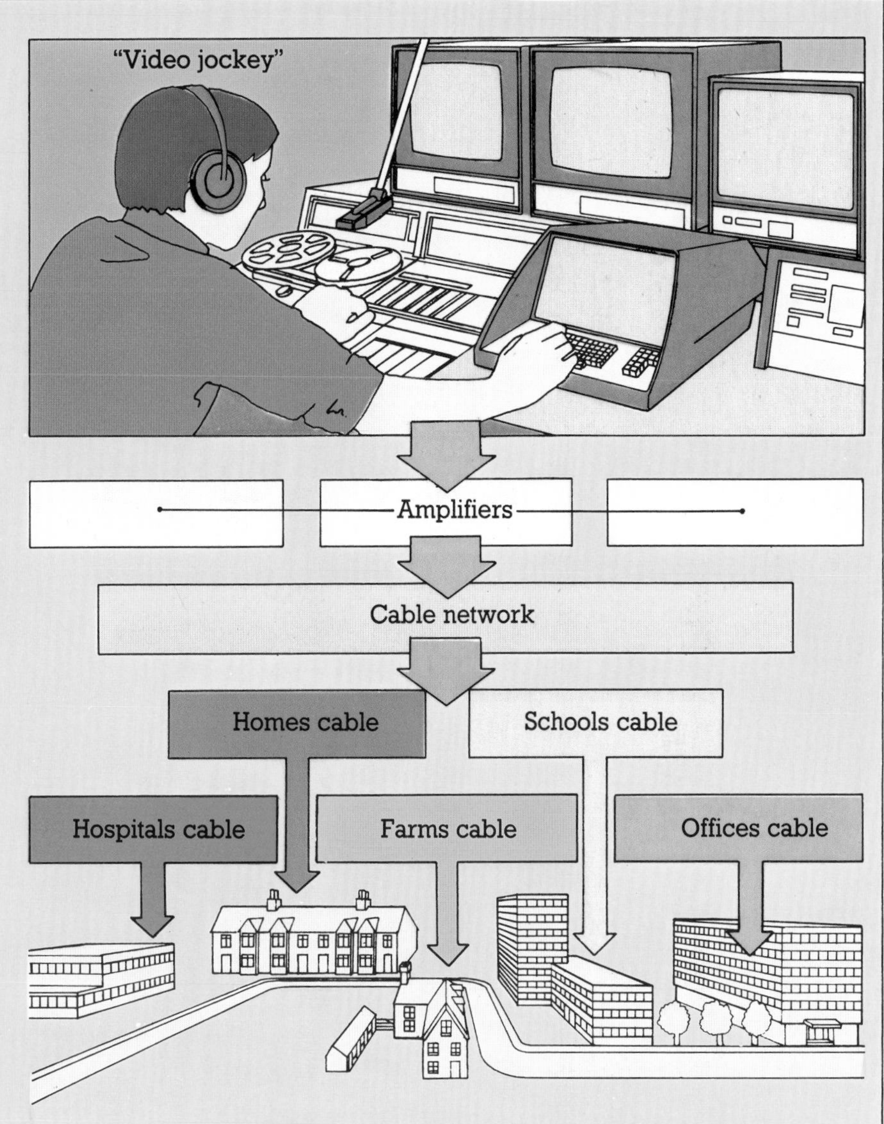

Cable networks can work hand-in-hand with broadcasting satellites. The satellite broadcast can be picked up by a number of local cable networks situated in the larger towns and cities. The cable network can relay the broadcast to subscribers in the city and others living in the outlying suburbs and rural areas.

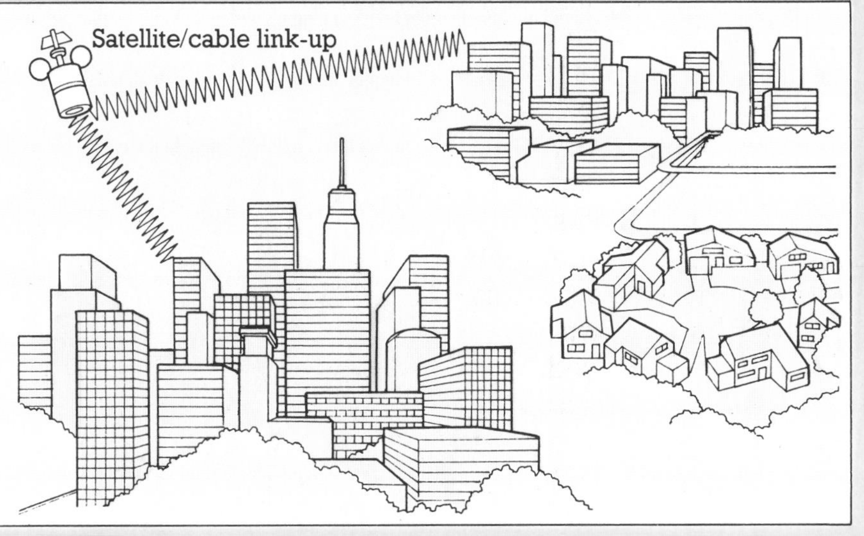

The Television Studio

As might be expected, it is in the television studio that the most intensive use of video equipment is found. Making a TV program – a week's episode of a serial, for example – requires a great deal of work from many people. Before the program gets anywhere near the studio, the director has to decide exactly what images he wants to show, and what video equipment he will need to achieve the result he wants.

When the writer has finished the script, the director will read it, and decide how it can be best turned into television. He may decide that certain scenes need close-ups of the action, but where will he place the camera? Will he need another camera taking a long-range shot? How will he cover three actors in one scene at the same time? Eventually, the director plans the whole program out into a series of shots, taken by different cameras in different positions, and he's ready to go into the studio.

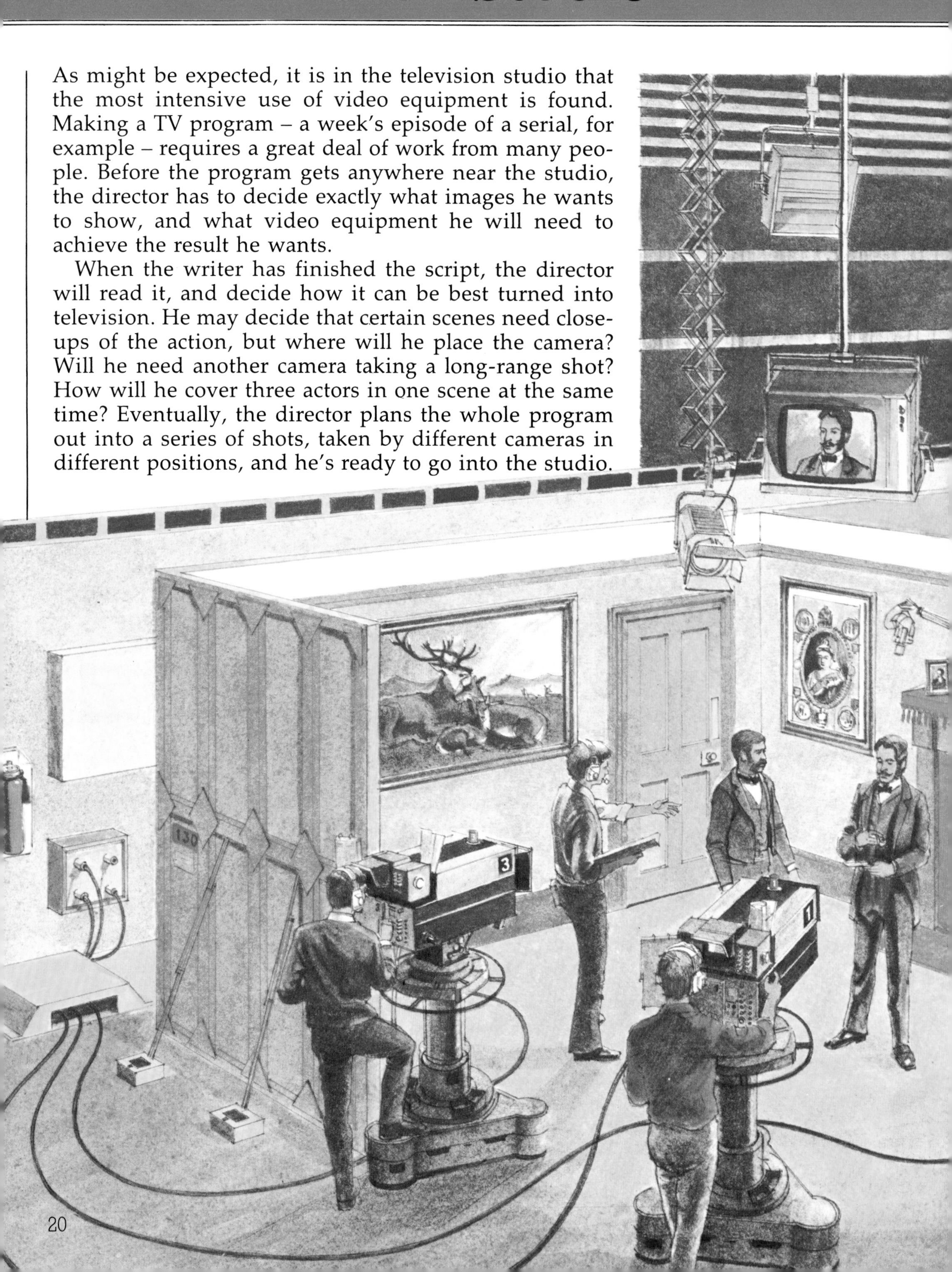

The studio will be dominated by the set for the program and by the glare of powerful studio lights that the video cameras need in order to record a good picture. There are video monitors instantaneously showing the pictures being taken by the cameras. Stretched out on long booms are the microphones to pick up the sound. When the director calls "Action" the tapes roll and the actors speak the lines they have been rehearsing.

The recording may go on for no more than a minute before the director calls "Cut!" to end that particular scene, and rearrange things for the next. Or perhaps something was wrong – the actors may have got their lines wrong or the camera may have caught the microphone in the top part of the picture – in which case the scene has to be shot again. But gradually, shot by shot, and scene by scene, the whole program is painstakingly pieced together.

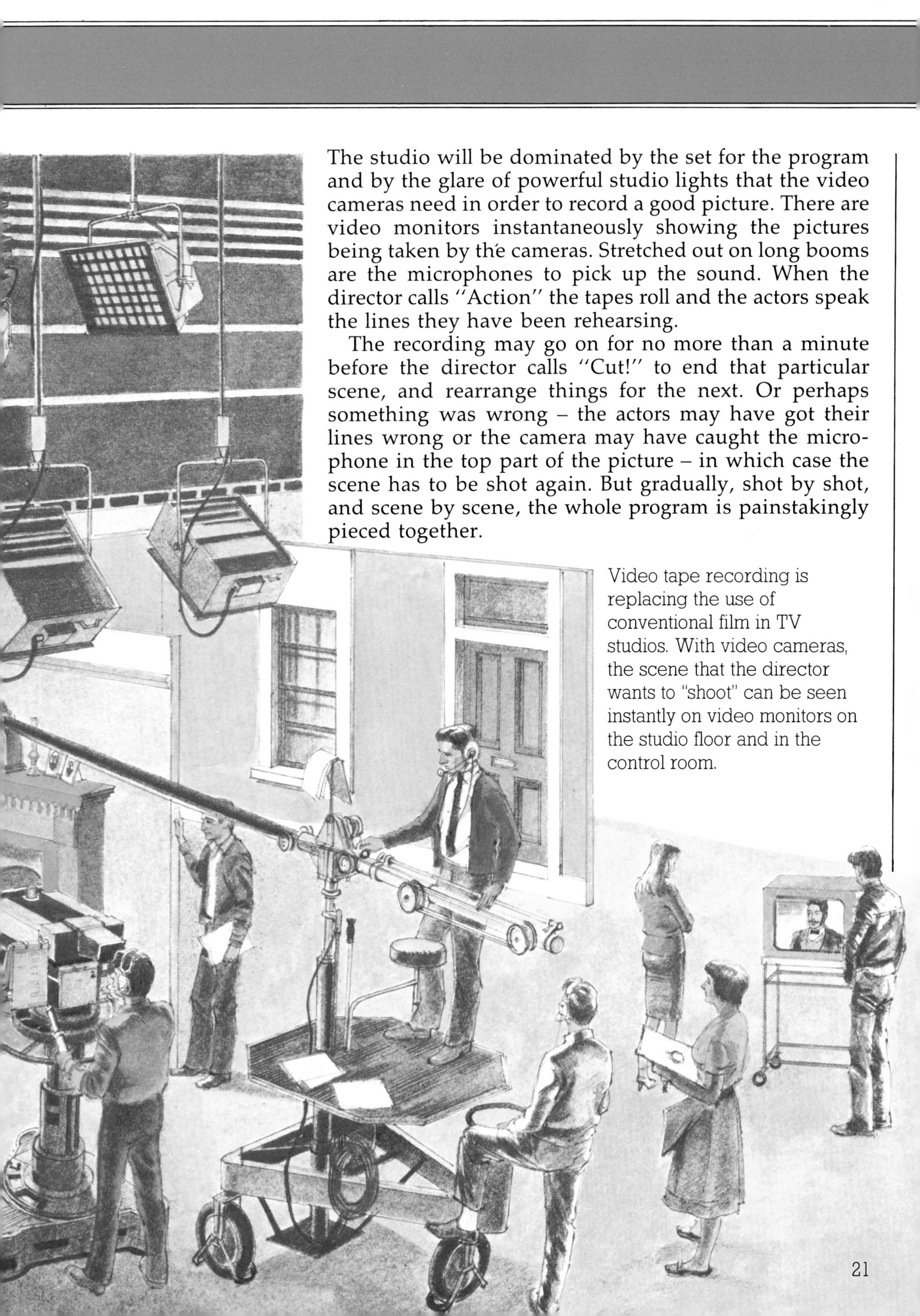

Video tape recording is replacing the use of conventional film in TV studios. With video cameras, the scene that the director wants to "shoot" can be seen instantly on video monitors on the studio floor and in the control room.

From the control room above the studio floor, the director can see what is going on in front of the cameras. There is also a bank of video monitors ranged in front of him, each showing the picture being seen by each of the different cameras. At the same time, he can listen to the sound as it is being recorded. A microphone link with the studio technicians enables him to give instructions to the studio floor. The scene illustrated above is a rehearsal, so the director will be watching to see if the shots he has worked out beforehand look just as he had expected them to. The white lines placed on the studio floor act as guides for the actors' movements, and will be removed when the actual shooting starts. In "live" broadcasts, the director chooses which of the different camera shots is going out "on air" at any given time. To make sure that the program runs smoothly and to its allotted time, he also gives instructions to the camera crew and the TV presenter: "Camera three, move in. Cue in camera one. Two minutes to go to the end of the show. Camera two on audience, please."

TV studio rehearsal

TV studio control room

Outside

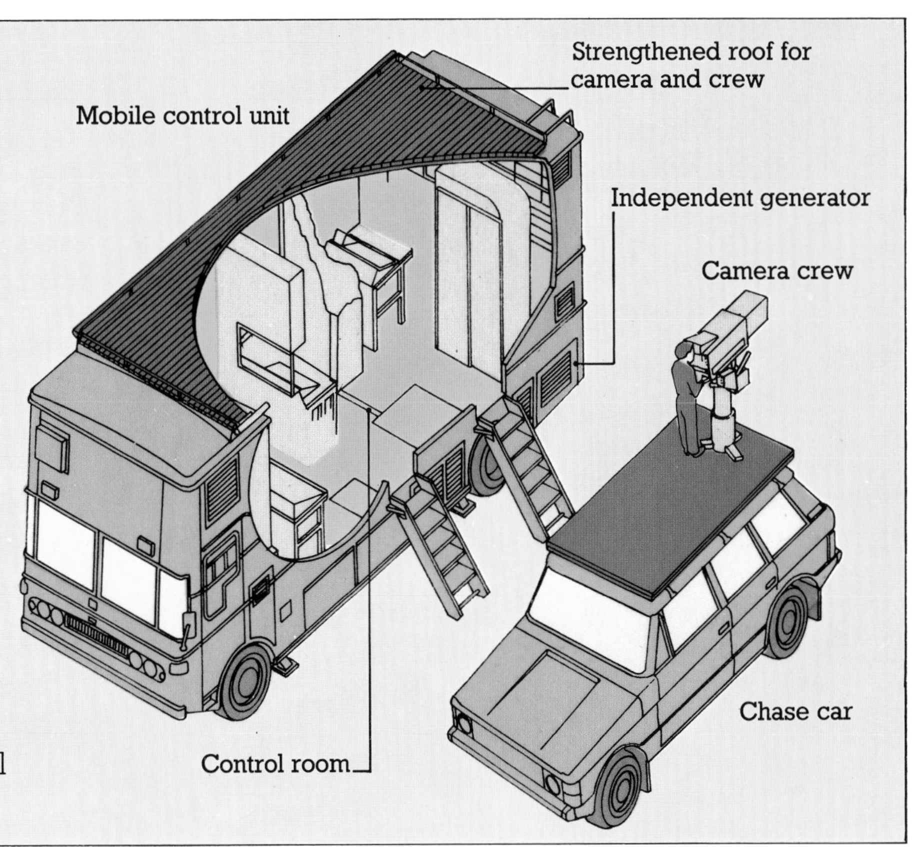

For outside broadcasts TV companies use a mobile control room, mounted on a truck. To cover a parade, for example, the producer has cameras at strategic points along the route. A "chase car" may also be used to follow the action. The cameras feed their pictures to video monitors in the mobile control unit, where the director chooses which will be broadcast.

E.N.G.

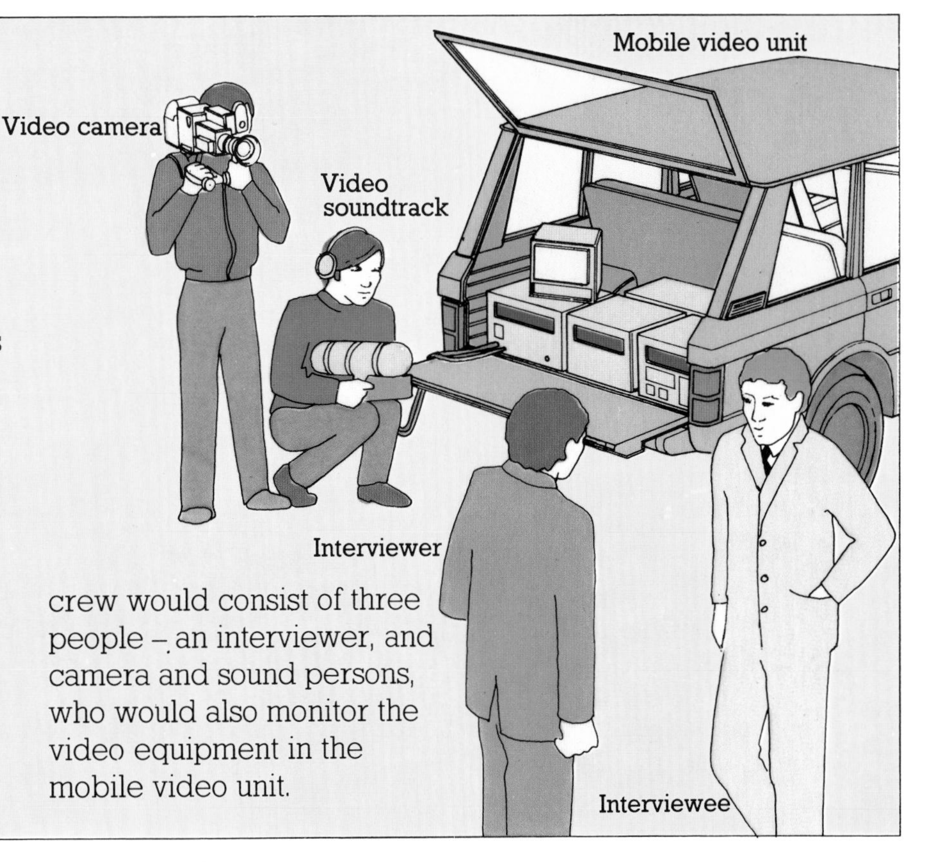

E.N.G. (Electronic News Gathering) is one way that the use of video techniques is affecting everyday television. Using video cameras, a news crew can be much smaller than an outside broadcast crew. And because the pictures are recorded on magnetic video tape, they can be viewed immediately, without having to wait for them to be processed in a film lab. A typical E.N.G. crew would consist of three people – an interviewer, and camera and sound persons, who would also monitor the video equipment in the mobile video unit.

Special Effects

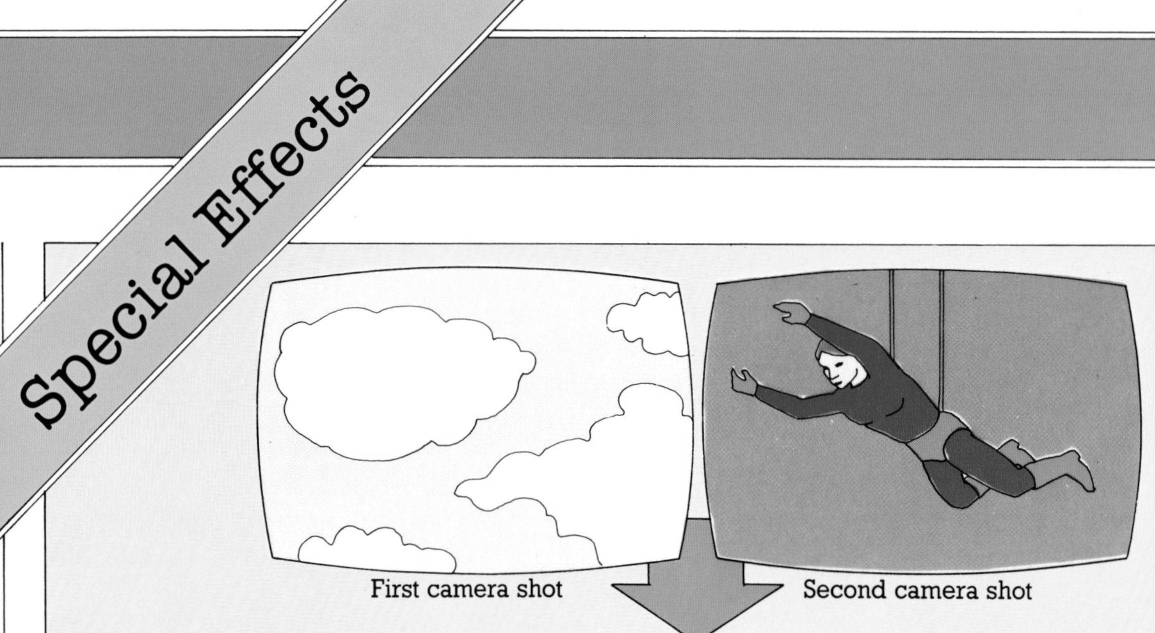

Combination of both shots

Many of the special effects seen in films, or on television, are created using a technique known as color separation overlay, or "chromakey." The method works by placing one picture on top of another one. To make an actor "fly," for example, the director would first take a shot of the sky. Then another shot of the actor suspended as if flying is taken. This shot needs careful planning. The background for the actor, and the wires holding him, are painted one particular color – blue, for example – and care is taken to make sure that there is none of that color in the actor's costume. This picture is then electronically processed to remove all of the background color from the picture. This leaves the actor against a completely black background. The first shot of the sky is then electronically "dropped in" behind him, to give a final picture of the actor flying through the air. This method can be used to create special effects for all kinds of programs.

News program on TV screen

Chromakey is used in news broadcasts, too. The pictures that appear behind the newscaster are electronically "dropped in." In the studio, all that can be seen is a blank card on the newsroom wall.

News program and chromakey panel from studio

Editing

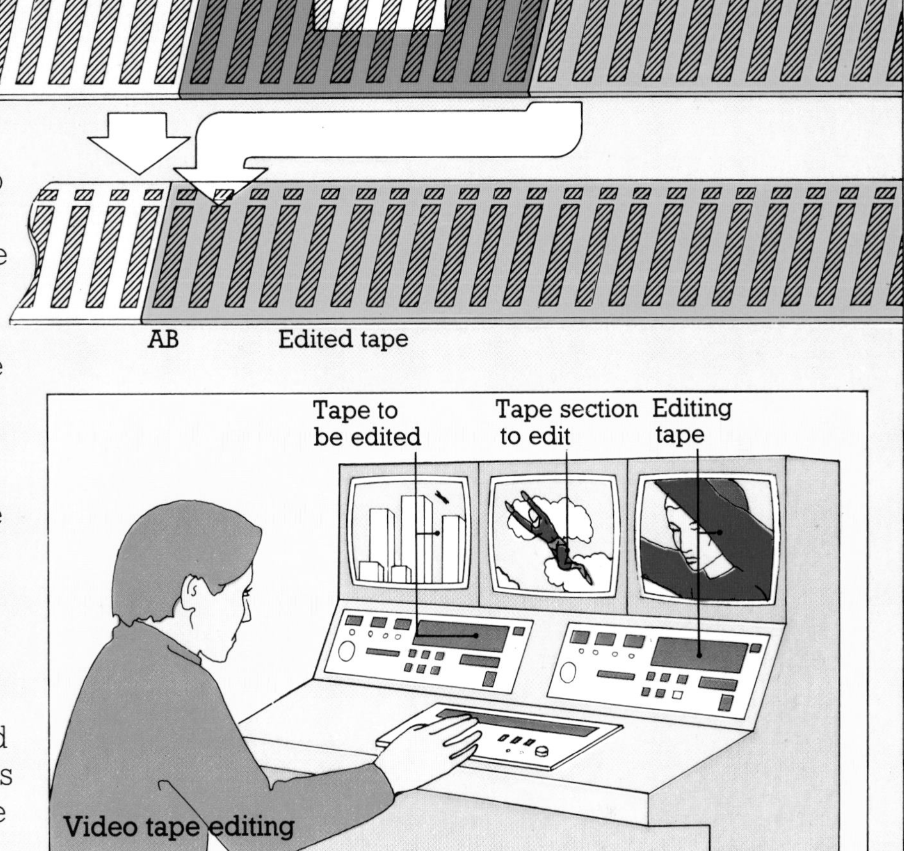

Video tape editing

To edit a program on video tape – to get it to the correct length, and to make sure that the story runs smoothly – two video recorders are needed. The first plays the video tape to be edited, and the second records the parts of the tape wanted. To cut out the section from A to B in the top tape, the second recorder records the tape up to point A, and is then switched off. The top tape runs through to point B, and the second recorder begins recording again to produce the edited tape.

Telecine

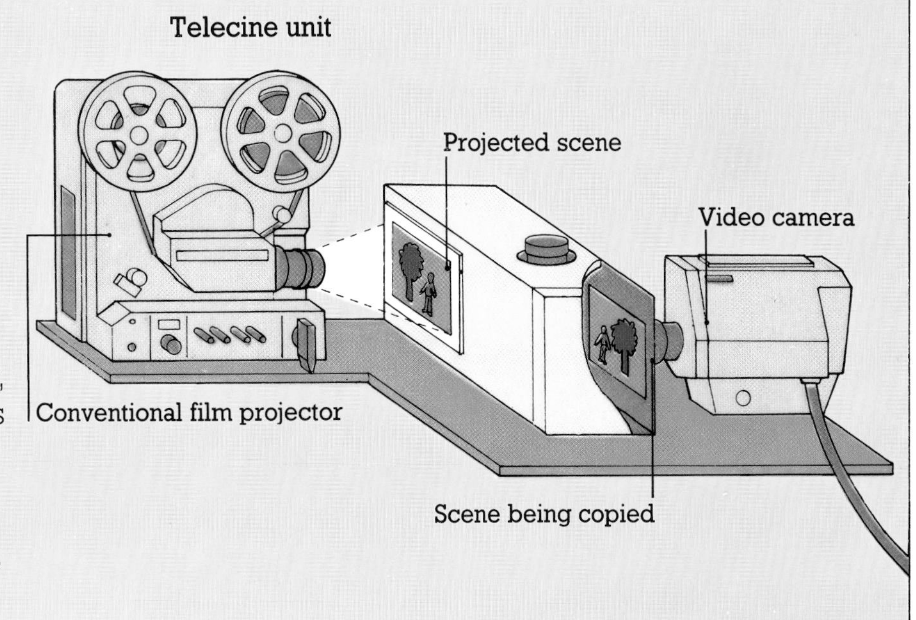

This "telecine" device is used to transfer movies from film to video tape. It projects the film into a box, and a video camera makes a recording of this projection. With telecine, you can also add a soundtrack to your favorite home movies.

The Flight Simulator

Just as he is coming in to land at London's Heathrow Airport, the pilot of this Boeing 767 jetliner feels a sudden lurch to the right, and a warning light flashes to tell him that he's lost power in his starboard engine. He struggles with the control column to get back onto the correct landing path, as the runway lights veer away from him. In fact, there is no danger, because the whole emergency is taking place in a flight simulator. But such is the accuracy of today's simulators, that it is easy for a pilot to forget that his "aircraft" is firmly bolted to the ground.

In effect, the pilot is playing a video game, but a far more complicated one than Space Invaders. The simulator is a full-size mock-up of the actual jetliner cockpit, with a number of large computers which generate images of the runway on the video screens that form the cockpit's windows, and give "real" readings on the instrument panel. The instructor sits behind the flight deck, telling the computers to create the various problems that the pilot has to deal with, either at take-off, during the flight, or when landing. The computers are so powerful that they can reproduce the runway approach to almost any airport in the world, and can simulate all types of weather and traffic conditions on the video screens. Using simulators, pilots can get a thorough training at a fraction of the cost of actual flying, and in complete safety.

Boeing 767 flight simulator

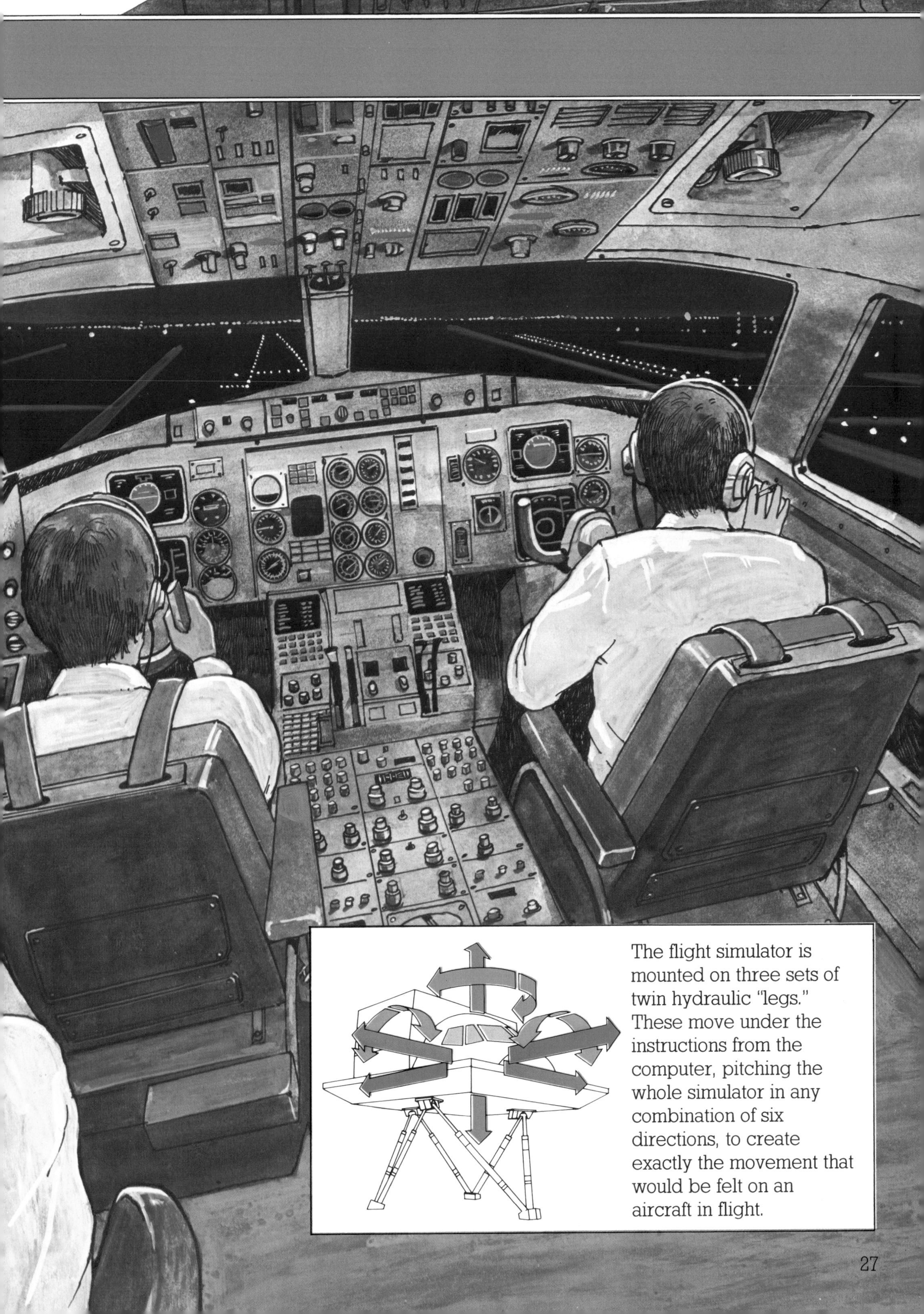

The flight simulator is mounted on three sets of twin hydraulic "legs." These move under the instructions from the computer, pitching the whole simulator in any combination of six directions, to create exactly the movement that would be felt on an aircraft in flight.

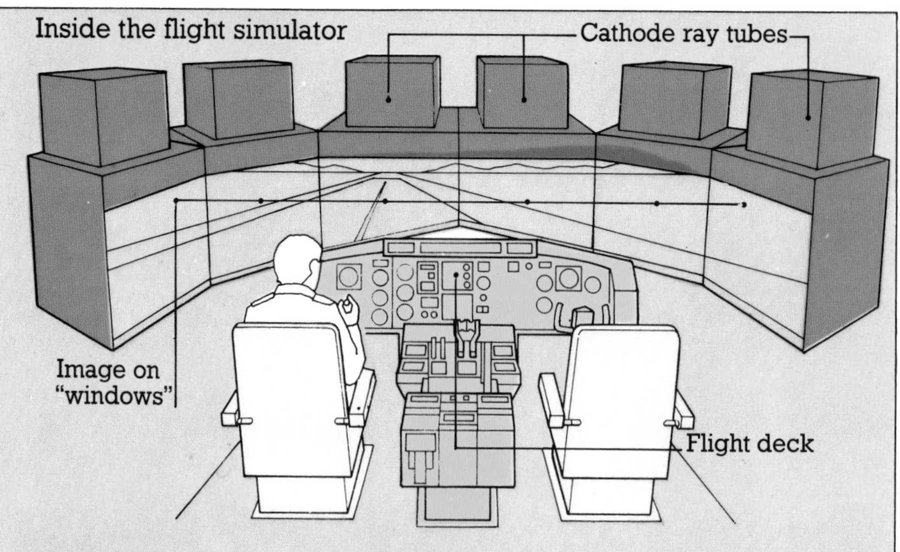

In the simulator, the image that the pilot sees from the "windows" is created on six TV screens. To create the image, the correct program is selected on the host computer (1). It might, for example, be asked to create New York's Kennedy Airport. The host computer sends the necessary information to the image-generating computer (2). This is programmed to send electronic signals to the cathode ray tube (3), to create the correct image. A sense of depth is given by reflecting the image off a mirror (4). Flight problems are also selected on the host computer. And when the pilot operates the flight controls, the instructions he gives are passed through the system, so that he sees the correct corresponding image in front of him. Early simulators could create only nighttime scenes, but the latest models create impressively realistic daylight scenes, like the two shown below. This enables the computers to create an even greater range of potential hazards for the pilot to cope with. The illustration on the right shows snow plows in the path of an incoming aircraft.

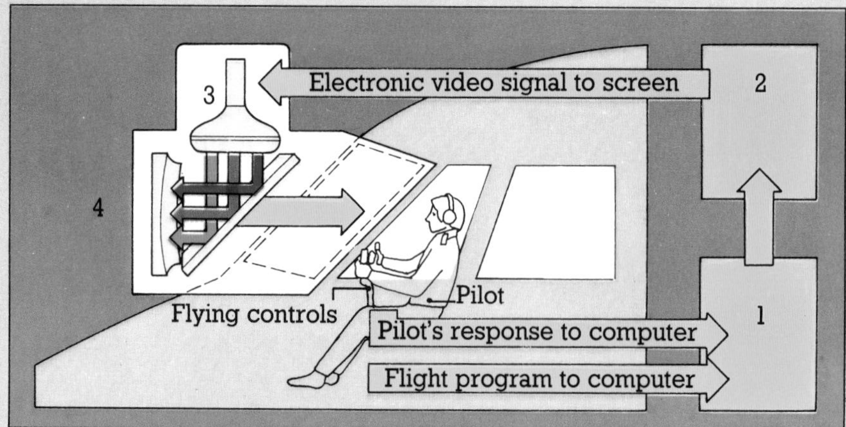

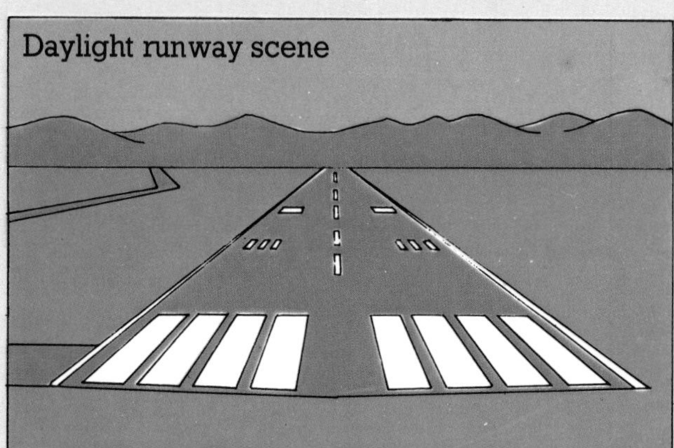

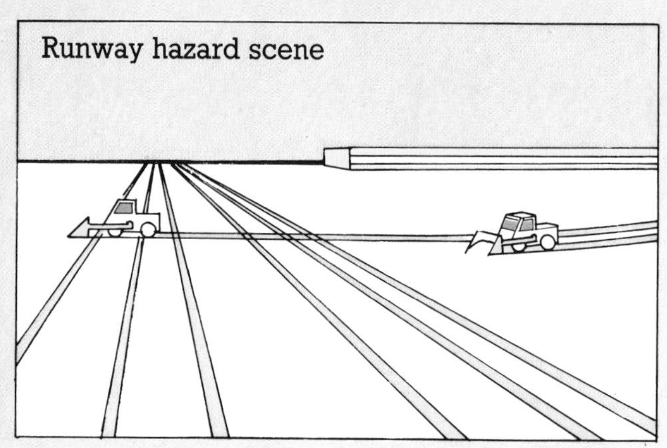

Seeing with Video

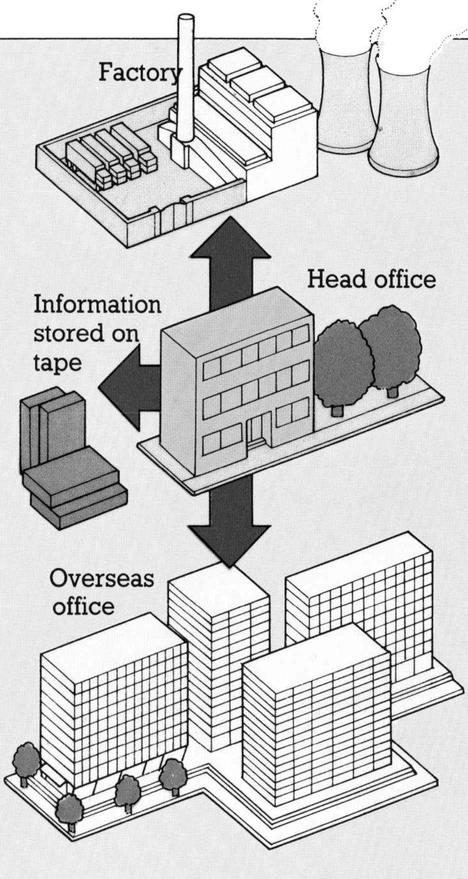

Video enables people to see and control things in a way that would otherwise be impossible. For example, a company's board meeting in one country can be relayed to staff in another, using satellite links, or it can be recorded on video tape for storage and future reference. In medicine (1), surgeons often use an "endoscope" – a flexible fiber optic tube that is introduced inside a patient. This gives them a closeup view of interior organs, which can either be seen direct or displayed on a video screen. Engineers in nuclear power plants (2), can see what's happening to the highly radioactive fuel rods, without risk to themselves, using video monitors. And video cameras can monitor production on robot assembly lines in the car industry (3). In many countries, video helps the police to keep a constant watch on the traffic using a minimum of manpower (4). In all kinds of areas today, video is proving to be an essential tool.

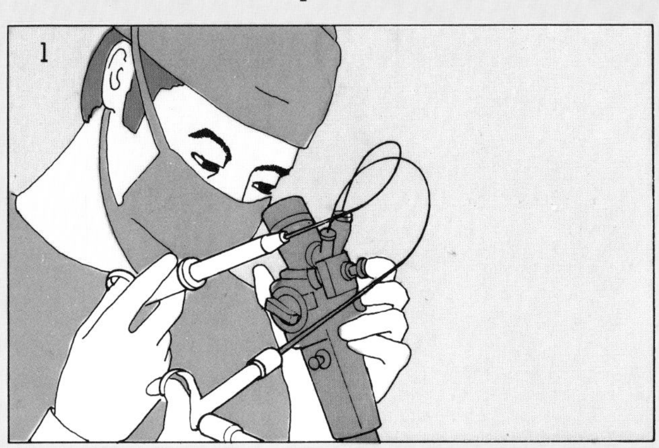

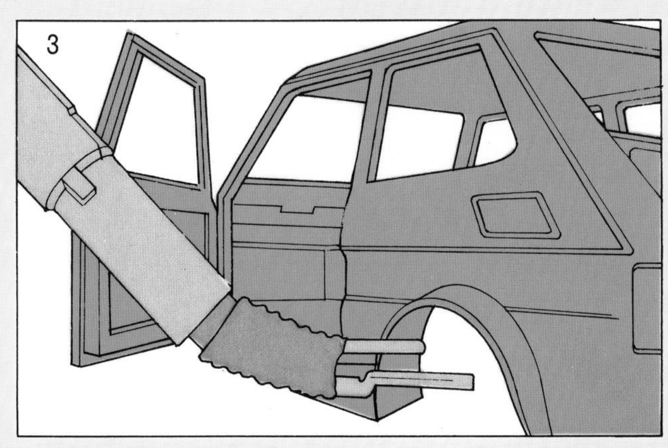

Video Information

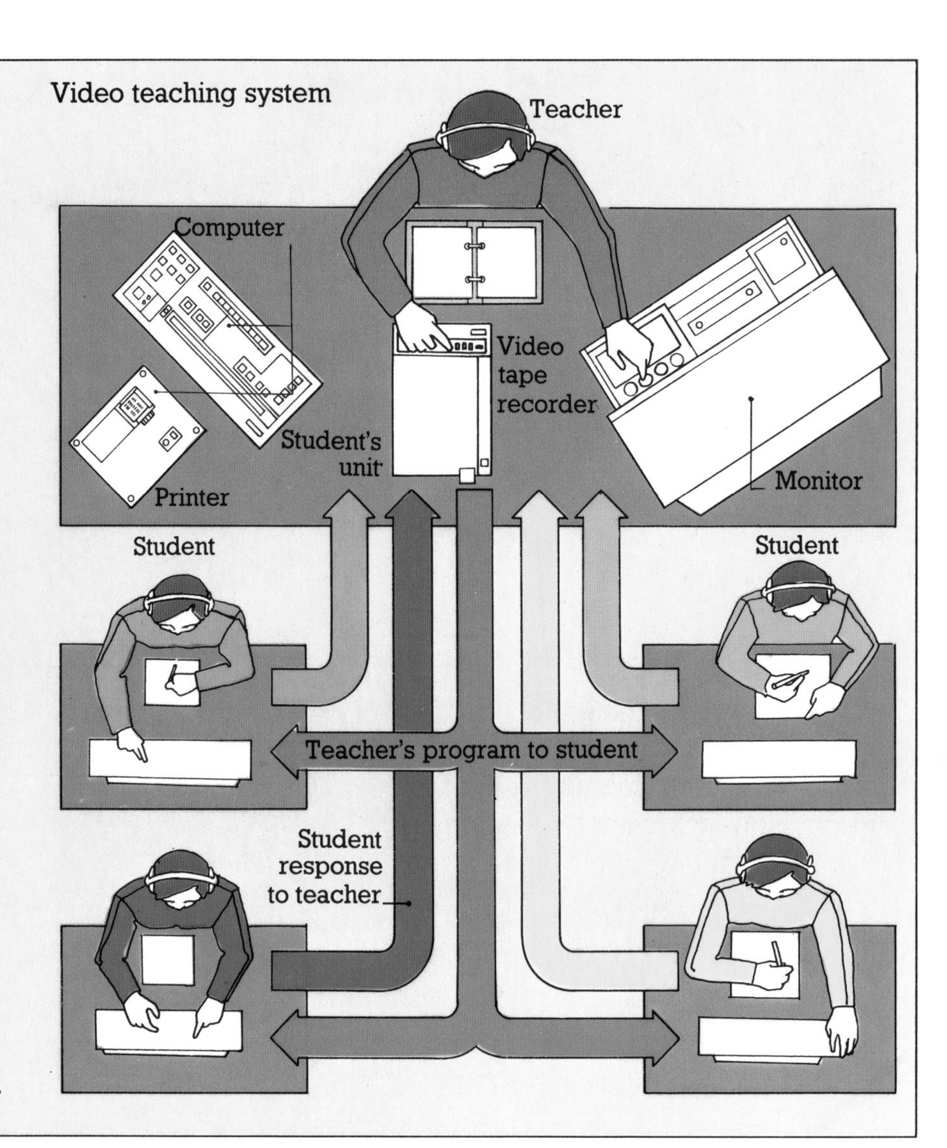

Video is extensively used as an educational aid in schools, colleges and industry. Video recorders can tape TV programs, or replay specially made educational tapes. The system shown opposite is particularly flexible. It uses a small computer to help students through a prepared sequence of video lessons. After watching part of the lesson, the students are asked questions by the computer. If they answer correctly, the computer moves on to the next part. If not, the lesson is replayed until the student understands it fully.

Cameras

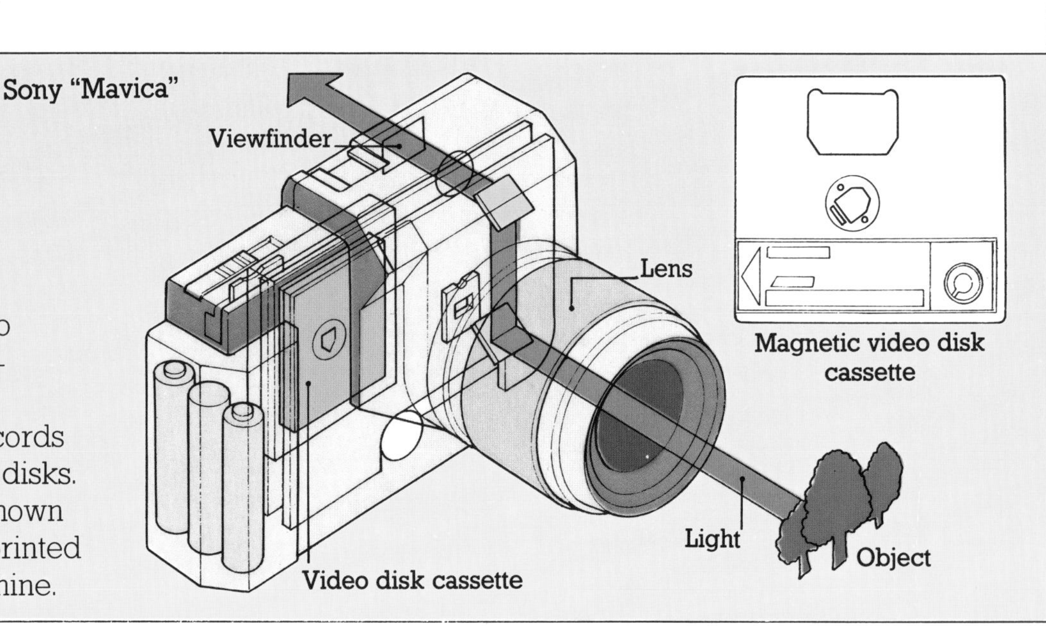

Video can be used to create still pictures – snapshots. The Sony "Mavica" camera records images on magnetic disks. The image can be shown on a TV screen, or printed using a special machine.

Viewdata

Computer information access

Information source

Adaptor unit

Host computer

Office monitor

Home monitor/TV receiver

Viewdata systems, like teletext, are really computers with general public access. Information from all over the world – airline schedules, for instance, or the price of shares on the Paris stock market – is fed to the computer via a telephone link. The users at the other end of the system, either in homes or offices, call up the information they want by pressing any one of over 200,000 "page" numbers on a special adaptor, linked to their own telephones. It's even possible to order merchandise, or write electronic letters to other users.

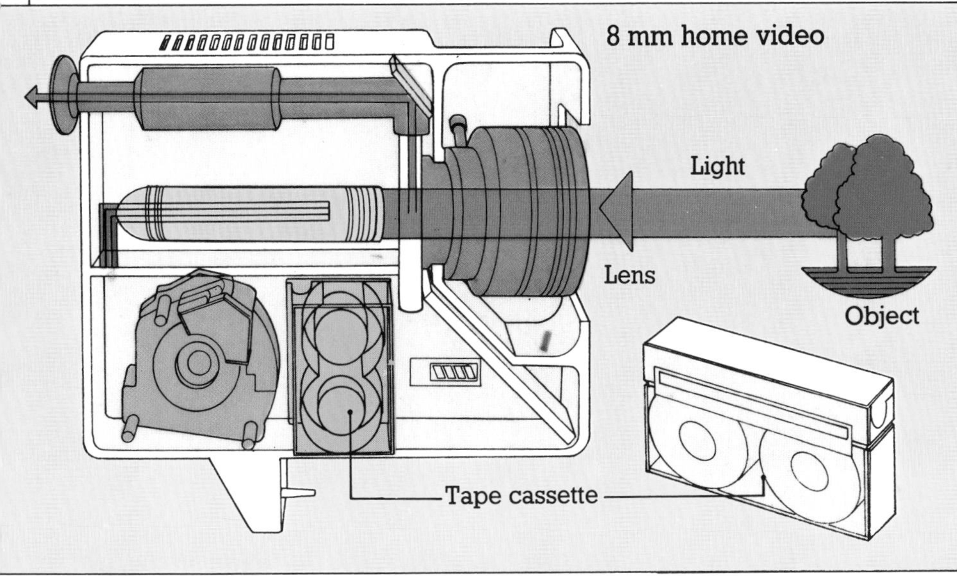

Video is already taking over in the field of home movies, and in a few years' time it could replace conventional film completely. Companies all over the world are working on video recorders that can be built into a video camera. Such cameras will be easy to operate and carry, and will use cassettes of video tape no bigger than a pocket diary.

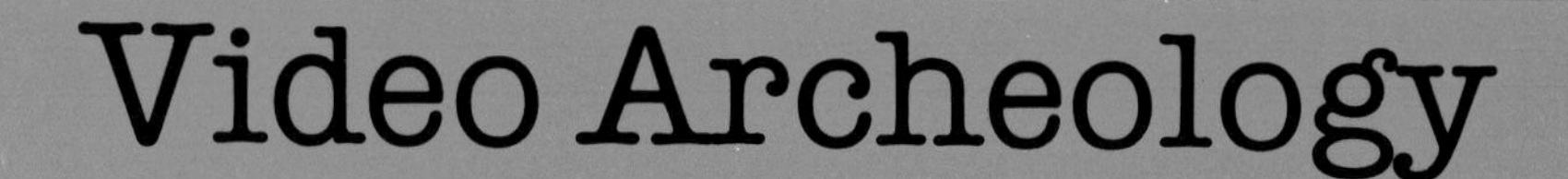

Video Archeology

One of the areas where video has proved most useful is in undersea work. Video cameras can operate at depths of more than 360 m (1,180 ft) – beyond the range of a diver. Video cameras mounted on special stands are used to monitor corrosion on undersea pipelines, and on the legs of oil rigs. But perhaps the most surprising undersea use of video is connected with an English Tudor warship – the *Mary Rose*.

The *Mary Rose* sank in the Solent channel, off Portsmouth, England, on July 19, 1545. But the fine layer of mud at the bottom of the Solent, up to 6 m (20 ft) deep in places, has kept her very well preserved. A team of archeologists and divers have been exploring the wreck, and it was finally brought to the surface in October 1982.

The muddy waters of the Solent mean that visibility was poor around the *Mary Rose*. In these conditions, a video camera can see about 15 percent better than the human eye, as well as giving the archeologists a permanent record of the excavation. As various bits and pieces were found within the wreck – items such as guns, surgeon's instruments, pottery, coins and even clothes – the exact position of each find was recorded on video tape. Using a special computer program, these video recordings will help archeologists to build up an exact picture of the *Mary Rose*, just as she was at the moment she was sunk, over 400 years ago.

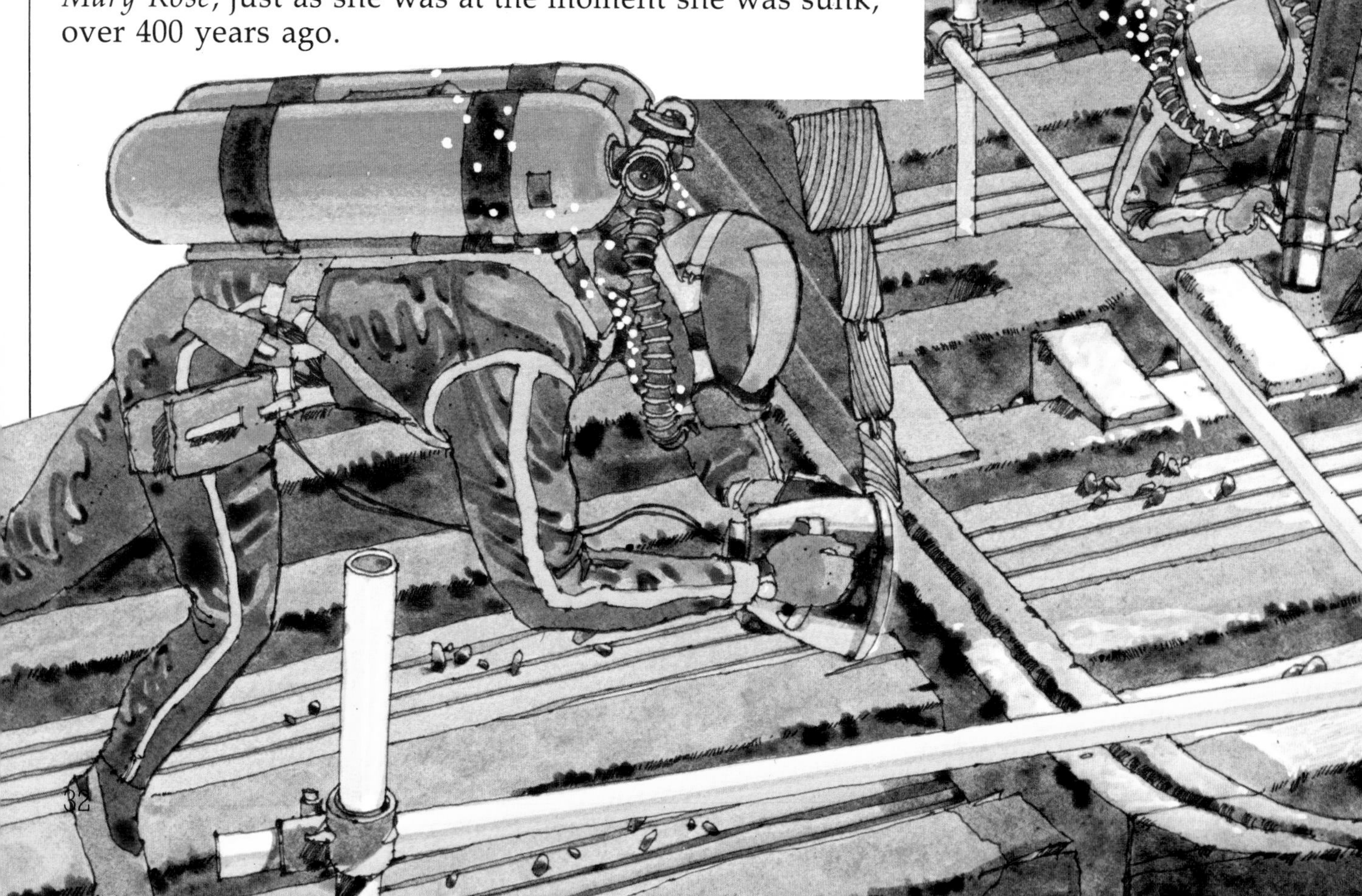

Archeologist in surface control room

The Image Intensifier

Special cameras, fitted with image intensifiers, can see things even when there is not enough light for the human eye. These cameras pick up the small amounts of light given off by an object, and amplify them until the image can be seen. Scientists and the armed forces in particular use these cameras.

Video Art

The images created by computers can have many uses. Engineers use them in technical design (1); graphs on video screens are used in business and science (2). The latest computers can also be used by artists to create their designs. The artist works at a special pad with a light pen connected to the computer. This recreates the lines drawn on the pad, on this video screen. The computer can fill in color as instructed, and also store every image the artist makes, for recall when required.

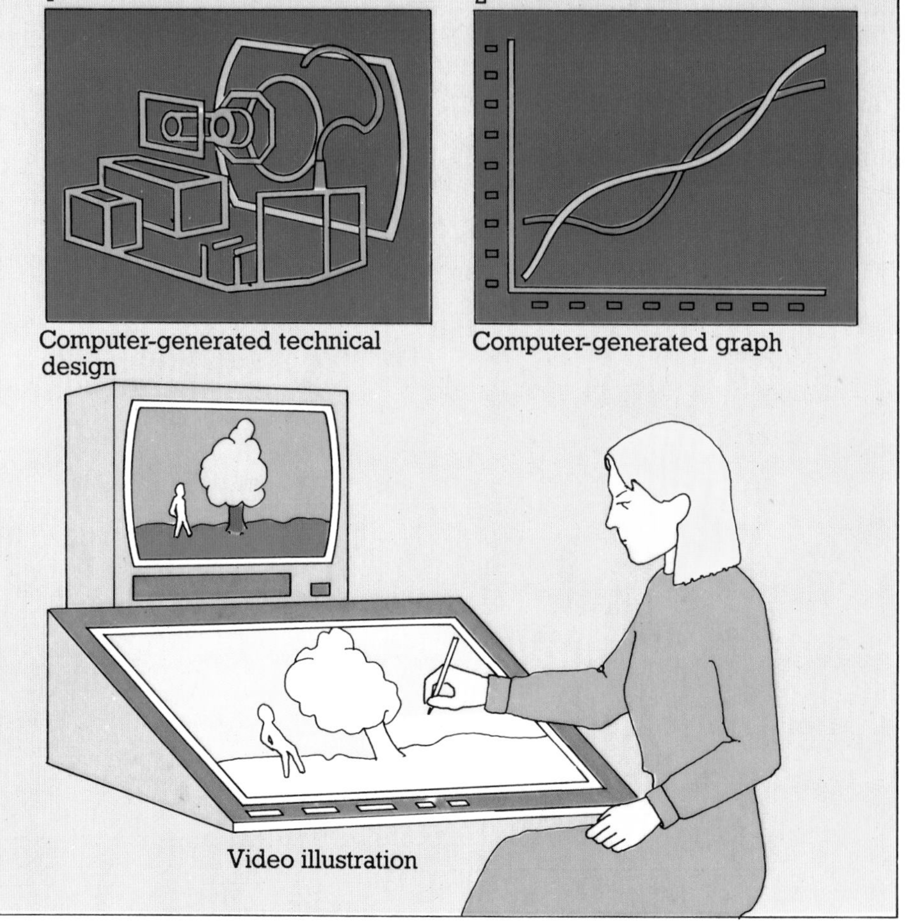

Computer-generated technical design

Computer-generated graph

Video illustration

A Video Future?

Today's video world is changing so rapidly that it is hard to predict the future. But within the next ten years video systems will be as familiar to us as ordinary TV is today. With the expansion of satellite and cable systems, the amount of information and entertainment available will be immense. Video will also make the world a smaller place, as we become able to talk to – and see – friends and colleagues living at the other side of the world by simply pressing a few buttons.

Factories will be almost fully automated, with robots doing the heavy, unpleasant or repetitious tasks, overseen by engineers using video cameras and monitors.

Traffic will be monitored by video cameras. With fewer people needing to travel to work, traffic congestion at peak hours in major cities will be less.

Shopping will be done by video. Customers will call up the store on their TV set, to find out current prices, place their order and have the goods delivered to their door.

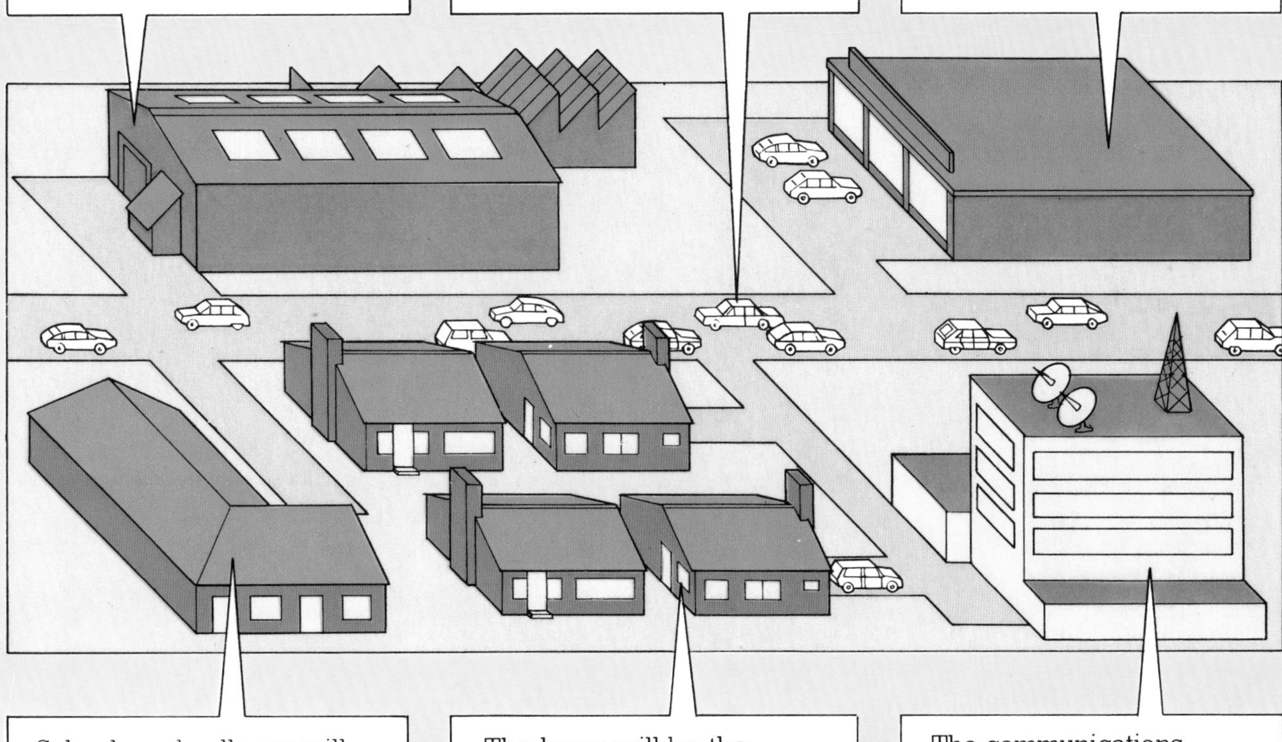

Schools and colleges will exploit the educational advantages of video and computers to the full. With video, students will "write" onto video tape rather than in exercise books.

The home will be the center of most people's lives, for both work and play. Video links to the office or factory mean that there will be no need to leave home to go to work.

The communications industry will become increasingly important in the future, linking up city to city and country to country in a worldwide information network.

Glossary

Cable network
The system of cables that links up all the cable subscribers to the central cable TV station.

Community TV
TV programs designed for a small, specialized audience, often for educational purposes.

Computer-generated images
Video pictures displayed on a TV screen that are actually created by the computer, rather than reproduced from drawings or photographs.

Conventional film
With conventional film, the light from an image creates chemical reactions on a thin layer of the film. Before the image can be seen, the film has to be treated in a special chemical bath.

Dish antenna
A large antenna, shaped much like a soup bowl. These antennas can pick up the relatively weak broadcasts sent down from satellites orbiting the Earth.

Editing
The process of selecting which particular images should be used in a film or TV broadcast.

Electrons
These are infinitesimally small pieces of matter, each of which carries the basic unit of negative electrical charge. An electric current is the flow of these particles along a conductor.

Electron guns
These give off a stream of electrons in the cathode ray tube when they receive high voltage electric current.

Fiber-optic cable
This is made out of very thin strands of special glass. Information is sent down them in the form of pulses of light, which are converted to an electrical signal at the receiving end. These are much more efficient and able to carry much more information than ordinary wire cables.

Host computer
The computer in a flight simulator which stores all the information needed for the simulator to work properly.

Laser beam
A concentrated beam of very pure light which can be precisely controlled and used to carry information.

Magnetic field
A field of force that is associated with certain metals, such as iron. A magnetic field is also always created by a moving electric current, and this is how it is created in the video recorder. The presence or absence of magnetic fields and their varying strengths, can be used to store information – as in the magnetic tape used in audio and video recorders.

Microcircuits
Very small electrical circuits printed onto "chips" of a substance called silicon. Incredibly, whole computers can now be put onto chips the size of a fingernail.

Phosphor dots
Phosphor is a substance which has the property of emitting light when struck by an electron. It is laid down in dots to create very small flashes of light that make up the TV picture.

Program
The special set of instructions that tells a computer what to do.

Solar cells
Devices which can convert the energy in sunlight into electrical energy.

Telecine
The process of converting film images into video images.

Transmitters
Very powerful electronic devices used to broadcast radio and television around the country.

Viewdata
A computer-based system for putting information onto video screens. The information can be displayed in the form of words or pictures.

Video head
The working part of a video recorder that lays down video information onto magnetic tape, and during replay, "reads" the video tape. Video heads are only a fraction of an inch in width.

Video information
All the instructions needed to build up a picture on a video screen.

Video monitors
Video screens that instantaneously display the pictures being taken by a video camera.

Video recorders
Machines which can record and replay video tapes.

Index

Aladdin Books would like to thank the following for their valuable help in the production of this book:

Sony (UK) Ltd, Grundig International Ltd, Thorn EMI Ferguson Ltd, Philips Electronics (Audio VLP Group), Matsushita Electric Corporation of America, JVC (UK) Ltd, RCA Columbia, Atari Inc, Marconi Avionics (Doug Howick), Prestel Public Relations, Video Information Centre, Ray Hodges Associates, Stanley Productions, BBC Photographic Library, Independent Television News Ltd, Thames Television Ltd, Tyne Tees Television Ltd, The Mary Rose Trust (Arthur Rogers) and Clive Gill.

Front endpapers: Various samples of video information (Grandfield Rork Collins & Partners), *back endpapers:* "LaserVision" video disk (Steve Sandon)

Printed in Belgium by Roost International Book Production

The Inside Story
VIDEO